Engineering Play

The John D. and Catherine T. MacArthur Foundation Series on Digital Media and Learning

Engineering Play: A Cultural History of Children's Software by Mizuko Ito

Hanging Out, Messing Around, and Geeking Out: Kids Living and Learning with New Media by Mizuko Ito, Sonja Baumer, Matteo Bittanti, danah boyd, Rachel Cody, Becky Herr-Stephenson, Heather A. Horst, Patricia G. Lange, Dilan Mahendran, Katynka Martínez, C. J. Pascoe, Dan Perkel, Laura Robinson, Christo Sims, Lisa Tripp, with contributions by Judd Antin, Megan Finn, Arthur Law, Annie Manion, Sarai Mitnick, Dan Schlossberg, and Sarita Yardi

Inaugural Series Volumes

These edited volumes were created through an interactive community review process and published online and in print in December 2007. They are the precursors to the peer-reviewed monographs in the series.

Civic Life Online: Learning How Digital Media Can Engage Youth, edited by W. Lance Bennett

Digital Media, Youth, and Credibility, edited by Miriam J. Metzger and Andrew J. Flanagin

Digital Youth, Innovation, and the Unexpected, edited by Tara McPherson

The Ecology of Games: Connecting Youth, Games, and Learning, edited by Katie Salen

Learning Race and Ethnicity: Youth and Digital Media, edited by Anna Everett

Youth, Identity, and Digital Media, edited by David Buckingham

ENGINEERING PLAY

A Cultural History of Children's Software

Mizuko Ito

The MIT Press
Cambridge, Massachusetts
London, England

For information about special quantity discounts, please email special_sales@ mitpress.mit.edu.

This book was set in Stone Sans and Stone Serif by SNP Best-set Typesetter Ltd., Hong Kong. Printed and bound in the United States of America.

Library of Congress Cataloging-in-Publication Data

Ito, Mizuko.
Engineering play : a cultural history of children's software / Mizuko Ito.
    p.   cm.—(The John D. and Catherine T. MacArthur Foundation series on digital media and learning)
Includes bibliographical references and index.
ISBN 978-0-262-01335-2 (hardcover : alk. paper)
1. Children's software—Development—History—United States 2. Computer-assisted instruction—United States. 3. Computers and children—United States. 4. Entertainment computing—United States. I. Title.
QA76.76.C54I86   2009
790.20285—dc22
                                                                                     2009006116

10 9 8 7 6 5 4 3 2 1

To Momoko Ito
(1939–1995)

Contents

Series Foreword

In recent years, digital media and networks have become embedded in our everyday lives and are part of broad-based changes to how we engage in knowledge production, communication, and creative expression. Unlike the early years in the development of computers and computer-based media, digital media are now *commonplace* and *pervasive*, having been taken up by a wide range of individuals and institutions in all walks of life. Digital media have escaped the boundaries of professional and formal practice, and the academic, governmental, and industry homes that initially fostered their development. Now they have been taken up by diverse populations and noninstitutionalized practices, including the peer activities of youth. Although specific forms of technology uptake are highly diverse, a generation is growing up in an era where digital media are part of the taken-for-granted social and cultural fabric of learning, play, and social communication.

This book series is founded upon the working hypothesis that those immersed in new digital tools and networks are engaged in an unprecedented exploration of language, games, social interaction, problem solving, and self-directed activity that leads to diverse forms of learning. These diverse forms of learning are reflected in expressions of identity, how individuals express independence and creativity, and in their ability to learn, exercise judgment, and think systematically.

The defining frame for this series is not a particular theoretical or disciplinary approach, nor is it a fixed set of topics. Rather, the series revolves around a constellation of topics investigated from multiple disciplinary and practical frames. The series as a whole looks at the relation between youth, learning, and digital media, but each might deal with only a subset of this constellation. Erecting strict topical boundaries can exclude some of the most important work in the field. For example, restricting the content of the series only to people of a certain age means artificially reifying an age boundary when the phenomenon demands otherwise. This becomes particularly problematic with new forms of online participation where one important outcome is the mixing of participants of different ages. The same goes for digital media, which are increasingly inseparable from analog and earlier media forms.

The series responds to certain changes in our media ecology that have important implications for learning. Specifically, these are new forms of media *literacy* and changes in the modes of media *participation*. Digital media are part of a convergence between interactive media (most notably gaming), online networks, and existing media forms. Navigating this media ecology involves a palette of literacies that are being defined through practice but require more scholarly scrutiny before they can be fully incorporated pervasively into educational initiatives. Media literacy involves not only ways of understanding, interpreting, and critiquing media, but also the means for creative and social expression, online search and navigation, and a host of new technical skills. The potential gap in literacies and participation skills creates new challenges for educators who struggle to bridge media engagement inside and outside the classroom.

The John D. and Catherine T. MacArthur Foundation Series on Digital Media and Learning, published by the MIT Press, aims to close these gaps and provide innovative ways of thinking about and using new forms of knowledge production, communication, and creative expression.

Acknowledgments

The social debts behind this book are immense, spanning two doctoral dissertations and a book revision. The fieldwork for this project was part of a large collaborative research effort. My team and colleagues in the Fifth Dimension project were an indispensable source of intellectual, emotional, and material support. In my immediate research team, Michael Cole, Ray McDermott, Jim Greeno, Shelley Goldman, and Don Bremme provided leadership, and in the project Charla Baugh, Katherine Brown, Gary Geating, Dena Hysell, Anne Mathison, and Karen Fiegener were fellow travelers and key colleagues throughout the effort. Vanessa Gack was my closest collaborator in the day-today fieldwork in the Fifth Dimension, and I still dearly miss her presence. I am grateful to the East Palo Alto Stanford Summer Academy for supporting our local club and to all the kids, undergraduates, and staff involved in the Fifth Dimension project who tolerated endless ethnographic scrutiny. The Institute for Research on Learning also deserves special mention as the institutional home of much of this project, providing the intellectual, administrative, and infrastructural support vital to this effort.

This work would not have been possible without the support of the Mellon Foundation and the Russell Sage Foundation, which have helped fund the Fifth Dimension project and this research. The Spencer Foundation's dissertation fellows program supported my time for analysis and writing. I am also grateful to Doug Sery, Mel Goldsipe, and Katie Helke at the MIT Press for their support and patience in helping bring this book to fruition.

I am deeply appreciative of those in the technology design and children's software community who took time out of their busy schedules to speak with an interested outsider: Chris Blackwell, Jonathan Blossom, Donald Brenner, Gary Carlston, Claire Curtin, Larry Doyle, Marabeth Grahame,

Amy Jo Kim, Scott Kim, Rita Levinson, Ann McCormick, Collette Michaud, Bob Mohl, Margo Nanny, Robin Raskin, Elizabeth Russell, Shannon Tobin, Peggy Weil, Will Wright, and Michael Wyman. I also thank the parents who offered a different perspective on this work: the late Rich Gold, Marina la Palma, Eduardo Pelegri-Llopart, Elizabeth Slada, and Karen Van De Vanter.

I thank the members of my dissertation and exam committees in both the School of Education and the Department of Anthropology at Stanford University: Carol Delaney, Ray McDermott, Joan Fujimura, Purnima Mankekar, Sylvia Yanagisako, Shelley Goldman, Jim Greeno, and Terry Winograd. Various portions of the dissertations and this book have benefited from readings by and conversations with many thoughtful colleagues in addition to my official readers and project team: Susan Newman, Annette Adler, Julian Bleecker, Seth Chaiklin, Lynn Cherny, Marjorie Goodwin, Donna Haraway, Marianne Hedegaard, Stefan Helmreich, Natalie Jeremijenko, Jean Lave, the late Timothy Leary, Elizabeth Mynatt, Vicki O'Day, Katie Salen, Ellen Seiter, Brian Cantwell Smith, Reed Stevens, Lucy Suchman, Kathy Wilson, and the members of my dissertation writers group, organized by Miyako Inoue—Pamela Ballinger, Genevieve Bell, Vera Michalchik, Liliana Suarez Navaz, Sandra Razieli, Shari Seider, Nikolai Ssorin-Chaikov, and Nancie Vargas.

Special thanks are due to those nearest and dearest who lived through this project at so many levels: Joichi Ito, for being my lifelong mentor in all things fun and computational; and Scott S. Fisher, for just about everything, but especially for companionship on all levels, personal and professional, and for an unflappable confidence in me through all the peaks and valleys. Midway through this project, I was joined by my children, Luna Ito-Fisher and Eamon Ito-Fisher, who put everything in perspective.

A summary of the arguments in this book was published as "Education vs. Entertainment: A Cultural History of Children's Software," in *The Ecology of Games: Connecting Youth, Games, and Learning*, ed. Katie Salen, 89–116 (Cambridge, Mass.: MIT Press, 2007). Portions of chapter 2 were previously published as "Engineering Play: Children's Software and the Cultural Politics of Edutainment," *Discourse* 27 (2): 136–160. Portions of chapter 3 were previously published as "Mobilizing Fun in the Production and Consumption of Children's Software," *Annals of the American Academy of Political and Social Science* 597 (1): 82–102.

1 Introduction

The relationship between children and computers occupies a special place in the imagination of those of us inhabiting the United States in the early twenty-first century. We think of kids as having a natural affinity to computers, as "digital natives" growing up in a world already saturated with computational media (Palfrey and Gasser 2008; Prensky 2006; Tapscott 1998). At times, though, this affinity becomes a source of fear and suspicion as kids reach out to unfamiliar social worlds online (see Cassell and Cramer 2007; L. Edwards 2005) or become "addicted" to fast-paced or violent forms of computer games (see Gunter 1998; Kutner and Olson 2008). More often than not, however, we see computers as a necessary part of kids' everyday lives and as valuable tools in supporting learning. We worry about a digital divide that puts advanced uses of computers out of some kids' reach (Seiter 2005; Warschauer 2003), so we have pushed for new technology to be present in schools and homes to enrich our children's lives (see Buckingham 2007; Buckingham and Scanlon 2003; Cuban 2001) and been for the most part successful in doing so: computers are ubiquitous in U.S. schools and are present in almost all homes with children (Roberts, Foehr, and Rideout 2005). Just three decades ago, the computer was associated almost exclusively with research, military, and business uses, but today it is widely recognized as a "children's machine" (Papert 1993). This book tells a key part of the story of how this transformation came about, describing how kids, technologists, and educators co-constructed new genres of computer-mediated play for children in the intervening years.

In the late 1970s, the idea of consumer software designed for the education, entertainment, and empowerment of children was barely a glimmer in the eye of a few radical educators and technologists. Released in 1977, the Apple II was beginning to demonstrate the power of personal computing and programming for hobbyists and educators. Programmers in a home-brewed industry began laying the foundations for a new consumer software industry by sending their products, floppy disks packaged in ziplock bags, to their networks of retailers and consumers. The video game industry hit public consciousness with the phenomenal success of *Space Invaders* in 1978, demonstrating the economic and addictive potential of interactive electronic entertainment. In tandem with these developments, small groups of educational researchers across the country began to experiment with personal computers (PCs) as a tool for creating interactive, child-driven, entertaining, and open-ended learning experiences. The trend toward a more child-centered approach to education and child rearing found material form in these technologies that allowed greater player control and input than traditional classroom media. Across a set of diverse contexts, educators and technologists began incubating a shared cultural imaginary that centered on new learning technologies.

Through the 1980s and 1990s, this experimental effort in software development blossomed into a new educational movement and eventually into a small but influential technology industry. Educators began to embrace these new technologies as alternatives to existing classroom computing, and in homes, learning games became a wholesome family entertainment alternative for young children. This period saw the emergence of a new category of consumer software designed specifically for elementary-age children, which blended different philosophies of education with genres and technologies drawn from interactive gaming and entertainment. Educators and technology designers experimented in creating a new set of media genres and a commercial sector that has variously been called *children's software, learning games,* and *edutainment.* Commercial children's software, designed to be both fun and enriching, lies in the boundary zone between education and entertainment that structures contemporary childhoods in the United States. The history of how children's software emerged as an experimental media category and its subsequent uptake by various social and political actors—including kids, parents, educators, and various commercial enterprises—is a microcosm of the social and cultural contestations surrounding new technology, children, and education. It describes efforts to incorporate gaming idioms into learning agendas and outlines the different understandings of learning and play that motivated these efforts. It is a story about the promise of technical innovation to transform the conditions of learning and play as well as a cautionary tale about the difficulties of reforming existing social and cultural structures, even with the best of intentions and innovative new technologies.

This book tells this story by describing how this new industry developed certain genres of software and how these genres were reproduced and challenged in the everyday play of kids in an after-school computer club where I conducted ethnographic fieldwork in the late 1990s. Its chapters are organized based on three genres of children's software—the *academic, entertainment,* and *construction* genres—that have manifested themselves in technology design and industry structure, as well as in *genres of participation* with and around the new media forms. The software genres grew out of existing educational approaches and everyday play in our culture at large. Where academic instructional software is based in a primarily behaviorist frame that focuses on the transmission of school-centered content, the entertainment genre is tied to open-ended play that is characteristic of family-friendly entertainment. The third genre is tied to constructivist and constructionist educational approaches that stress authoring and media production as a vehicle of learning. All three are malleable, evolving, but also recognizable patterns in our culture that take shape because they are deeply embedded in institutions and social practice. New technology practice grows out of existing genres, but also reshapes them in important ways.

The book's primary descriptive task is to delineate these genres of children's software in a way that cuts across the usual distinctions we make between moments of production, distribution, marketing, and consumption. My focus is on analyzing how the positioning of and the contestation between different genres and educational approaches are embedded within a broader systemic and historical context. The story I tell is not of an inevitable realization of technological potential, but rather of ongoing struggle, negotiation, and contestation between different actors and social and cultural forces. This work aims to inform contemporary debates in education over games and learning by looking back to the lessons we have learned from an earlier generation of learning-software products. Although the book looks at particular historical moments in the evolution of software for children, the underlying genres I describe are very much still in play today.

This descriptive task is closely tied to my primary theoretical task, which is to identify the dynamics through which new media become incorporated into and in turn reshape existing social and cultural structure. The emergence of children's software and its subsequent incorporation into existing institutional and industry structures offer an illustration of the dynamics of technology development, diffusion, domestication, and appropriation. By looking not only at the content of software, but also at the broader social and cultural contexts in which the software is embedded, we can understand how it is that technology fails to deliver on some of the explicit claims made by its developers and at the same time is tied to broader systemic changes. I suggest that the issue is not one of technology innovation being incorporated into and stifled by existing discourses and institutions. Rather, new technologies never start out as separate or outside of these existing structures. Change happens as a result of struggles and negotiations *between* different discourses and institutions seeking to shape a new technology and set of genres. In the remainder of this introductory chapter, I describe the conceptual frameworks that I draw from in education and new media studies and then turn to a description of the fieldwork and sites where I conducted my research.

Games and Learning

Interactive media such as computer games, simulations, and digital author-ing tools added new impetus to longstanding efforts to integrate play and education in commercial media targeted to families. Educational toys, books, magazines, television, videos, and software have defined a market of products that cater to parents interested in enriching their children's home activities with media explicitly designed for learning. David Buckingham and Margaret Scanlon describe the growing market for these kinds of products as an indication of the "curricularization of family life" (2003, 6), where parents are called on to support educational goals at home. Looking to media as a source of education is tied to what Stewart M. Hoover, Lynn Shofield Clark, and Diane F. Alters (2004) have described as a stance of "reflexive parenting," where parents reflect on their home life in relation to public discourse and expert advice about parenting. Their work discusses "public scripts" about the effects of media on children and how parents feel pressure to manage and regulate media in the home.

These public scripts about media and children are constantly being redefined and contested, and new media are part of this ongoing process of collective negotiation. When PCs became a viable consumer technology, software developers and educators began developing new genres of software and new public scripts about computational media and learning. They argued that new interactive media held out the potential to challenge the dominance of "passive" forms of media that were exemplified by the television. The computer was defined as a "good screen" that was contrasted to the "bad screen" of the television (Seiter 1999) and that contributed to children's learning and enrichment. Whereas video games played on TV screens and in game arcades continue to be defined as problematic forms of media that parents should regulate and control, computers and computer software have occupied a privileged position in the public scripts about media and learning. Buckingham and Scanlon argue that "the home computer could be seen as one of the indispensable 'symbolic goods' of contemporary parenting" (2003, 109).

The question of what constitutes appropriate forms of software for children has evolved in tandem with changing technologies and our changing discourses about technology and learning. The category of learning software for children is a potentially broad one, and one can imagine multiple origin stories and histories. Here I focus on a particular trajectory with origins in the late 1970s and early 1980s, when educators and technology makers built a new industry niche of software products for elementary-age children. This period saw the founding of a category of software that came to be called *edutainment* or, more broadly, *children's software*. These software titles were offered as an alternative to the drill and curriculum-based, computer-aided instruction systems such as PLATO and Wicat that dominated the educational technology field from the 1960s to the 1980s. Instead, they drew from the video game culture being developed for arcades and game consoles and applied elements of that culture to educational technology. Games such as *Number Munchers, Oregon Trail, Reader Rabbit, KidPix,* and *Where in the World Is Carmen San Diego* were in this first major wave of software designed with learning goals in mind and targeted toward the consumer market of families with elementary-age children. In this approach, children's software drew from a longer history of educational reform efforts that looked to play as a site of learning.

Although not the only way in which playful idioms were incorporated into these software titles, the use of electronic gaming idioms is what was distinctive about this particular historical moment in the evolution of the philosophies of "learning through play." In contrast to earlier educative playthings and media—typified by wooden blocks, puzzles, children's literature, and *Sesame Street*—educational software put gaming at the center of the enterprise. By focusing on software and games explicitly designed to mediate between educational and entertainment idioms, I do not mean to privilege the learning claims made by these titles. Clearly, many titles in the mainstream entertainment gaming market embody important learning principles (Gee 2003; Ito 2006b). Rather, the importance of edutainment as a case is that it exemplifies the struggle to identify a cultural and social space between the polarities of education and entertainment that define modern childhoods. The boom and bust of the children's software industry from the 1980s to the early 2000s illustrate the cultural, social, and capitalist contestation over media for children as the industry intersected with changing platforms for digital entertainment.

This educational software movement was not monolithic; it was differentiated by distinct educational philosophies and design approaches (Engenfeldt-Nielsen 2006). I describe these differences in terms of the academic, entertainment, and construction genres of software and participation. These three strands of software development fed into and informed one another to build a new niche software industry. Each involved translating specific educational philosophies and representational genres into a new media platform. I identified these three strands by working backward from the ethnographic record. In the computer clubs where I conducted my fieldwork, what constitutes the general category "children's software" or "learning games" is remarkably consistent; publications aimed at teachers, educationally minded parents, and after-school program staff identify these games as educational and prosocial. I thus use the term *children's software* to refer to a category of commercial software that is targeted toward elementary-age children and that embodies general cultural commitments to learning and developmental goals. Educational or learning games can be considered a subset of this category, which also includes authoring and simulation tools that don't have a strong gaming component. In the description to follow, I sharpen the difference between the three media genres in order to identify competing cultural codes and educational philosophies embodied in children's software. In the real practices of play and in the design of specific titles, however, these genres are often intermingled. An academic title may embed elements from entertainment media or elements from authoring tools, just as an entertainment-oriented title may work to convey some curricular content.

This book describes a particular moment in the history of children's software and learning games, a period when new markets were developed and key genres were defined. It was a period in which computers became pervasive in children's lives, in schools, and in many homes in the United States as part of the growing consensus about the educative potential of computing. Although the technology and industry have evolved tremendously since the 1980s and 1990s, the genres, discourses, and industry structures established then are still very much in play today. Much of what we take for granted in the relationship between children, learning, and computers was a result of the struggles regarding the technology and its uses during that period. The negotiation between the educational and entertainment genres, between school and home, is a resilient feature of our social landscape and has been concretized through the software genres that this book describes. The academic genre has a niche in supporting and reinforcing certain basic literacy and mathematical skills in connection with school-based accountabilities. The entertainment genre, in opposition to the academic, is defined by a kid-centered and kid-identified visual and peer culture based on media in the home. The construction genre represents the most profound influence of computational media on learning in that it expands the range of opportunities for kids to author and reshape media worlds. In the conclusion, I return to how these genres are playing out in the current landscape of new media and learning.

The lessons from the early years of edutainment software can also inform current debates over games and learning. We are seeing a revival of arguments in favor of the learning benefits of game play as a new set of networked and technically sophisticated games and authoring tools have captured the attentions of a new generation. These arguments are wide ranging, but many replicate those put forth during the early years of edutainment, though with a much broader palette of application. Game developers working in the area of "serious games" or "games for training" have argued that games can cross over the boundaries of entertainment and established institutions and discourses of learning (Michael and Chen

2006; Prensky 2006). Mark Prensky describes what he sees as a "digital game-based learning revolution," which is transforming corporate and military training. "The days of sitting bored to tears in classrooms or in front of boring computer or Web-based training systems are numbered" (2001, 34). Just as early developers of edutainment worked to break down the barrier between the classroom and commercial video games, today's developers are working to integrate curricular and training goals with the interactive mechanisms and visual idioms of entertainment software. If our past experience is any guide, we can expect that new software and markets will emerge to cater to these different educational contexts, but that these efforts will be largely oriented to supporting existing institutional accountabilities.

The growth in the visibility of sophisticated new games is also tied to a burgeoning research literature on games and learning. Much of the current work has departed from the early paradigms that gave rise to edutainment. Most notably, recent research has challenged the assumption that only games with academic content have learning properties, and that gaming should be used to "sugar coat" otherwise unappealing learning tasks. In his influential book *What Video Games Have to Teach Us about Learning and Literacy,* James Paul Gee (2003) argues that successful video games, even those without explicitly educational goals, support sophisticated forms of literacy by situating learning within responsive, embodied, and challenging environments. Gee also stresses the socially and culturally situated nature of the learning that takes place through gaming. Building on Gee's work, Kurt Squire (2007) suggests that state-of-the-art video games provide rich opportunities for learning through immersive action and performance, and he critiques games such as *Math Blaster* that rely on game play as an extrinsic reward to get players to engage with the otherwise boring task of math drills. By contrast, he presents examples of classroom use of *Civilization III,* where the educational content is integrally related to the game-play experience and simulation, and where students are actively solving problems that push them to reflect on the dynamics of history and politics.

Complementing Gee and his colleagues' work, other researchers have argued that games encourage particular kinds of literacies and learning dispositions. For example, Ian Bogost (2007) contends that games are uniquely well equipped to educate players on "procedural literacy," which involves an understanding of the workings of various systems of relationships. He argues that computer games provide a rich set of building blocks that can be manipulated and recombined, whether in learning computer programming, tinkering with a simulation of city or civilization, or building a virtual home. David Shaffer (2006) likewise suggests that the unique properties of simulations make them powerful vehicles for learning. He argues for the learning benefits of "epistemic games" where players can take part in activities that simulate the practices of professions such as historians, engineers, and mathematicians. He sees computer-based games and simulations as powerful tools to "develop the skills, knowledge, identities, values, and epistemology of that community" (164). This set of research conversations is less focused on games as a platform to deliver academic content and more on how game play can support a wide range of learning outcomes, both social and cognitive.

Current research on games and learning has clearly moved beyond many of the conceptual paradigms that structured early educational games, and the field has been enriched and expanded in a way that would have been difficult to imagine in those initial years. My intention here is not to conduct a thorough review of the field of games and learning, which has blossomed in the period between when I did my fieldwork in the 1990s and the publication of this book. My hope, however, is that a look back to a prior period of technical innovation and cultural struggle can help inform current research approaches. The book contributes more to this field than simply providing a lesson regarding the failures and successes of an earlier generation of technical innovation and educational reform. In addition to this historical lesson, it makes an argument for a conceptual focus that challenges certain tendencies in the research literature. Although the field of game studies has made great strides in theorizing the unique structure and learning properties of games, there is still relatively little empirical research on practices of game play and virtually no work that situates this play within an analysis of broader institutional structures such as the marketing and distribution of software, the production of class distinctions, and the discourses of parenting. The focus on game content and design and the relative absence of work on more contextual dimensions reinforce a perception that the technology and media content itself are determining the learning outcomes. We need to do more than simply point to what kids can potentially get out of playing games, however, and consider how these games are embedded in their everyday lives and in institutional accountabilities. By failing to do so, we will continue to over-estimate the force of technology in transforming education and to under-estimate the role of institutions and existing practices in determining learning outcomes.

Some work is beginning to examine the contexts of game play, thus extending beyond a narrow focus on the relation between a player and game content.[1] Nevertheless, almost all empirical studies of gaming prac-tice and learning are focused on learning and literacy outcomes rather on broader trajectories of social and systemic change in which these outcomes are situated. Critical studies of gaming and edutainment industries do examine some of these broader dynamics in relation to the content of game software (Buckingham 2007; Buckingham and Scanlon 2003; Kline, Dyer-Witherford, and de Peuter 2003), but they do not bring the analysis down to the level of everyday gaming practice. This book is an effort to integrate these levels of analysis in that it combines detailed practice-based description of game-play practice with analysis of broader cultural dis-courses and social and institutional structures. It is only through an exami-nation of how game content, design, and play are located within social and cultural structures that we can begin to understand the broader out-comes of our efforts to shape and engineer learning and play. This gap in the literature is not merely an empirical gap, but is tied to a more funda-mental set of conceptual and theoretical dynamics that have characterized research on new media and social change. The remainder of this introduc-tion describes my conceptual framework for understanding the dynamics of structural conservatism and change, and then turns to a description of the methods and fieldwork behind this book.

Technology and Social Change

In any examination of new technology, it is a challenge to avoid the pitfalls of both hype and mistrust, or, as Sarah Holloway and Gill Valentine (2001) have described it, the problem of polarization between the "boosters" and the "debunkers." New technologies tend to be accompanied by a set of heightened expectations, but then are followed by a precipitous fall from grace after failing to deliver on an unrealistic billing. This was certainly the case with edutainment, which boosters hoped would transform learn-ing for a generation of kids. Although the boosters and debunkers may seem to operate under completely different frames of reference, they share the tendency to fetishize technology as a force with its own internal logic standing outside of history, society, and culture. The problem with both stances is that they fail to recognize that technologies are in fact embodiments, stabilizations, and concretizations of existing social structures and cultural meanings, and that they grow out of an unfolding history as part of a necessarily altered and contested future. The promises and the pitfalls of certain technological forms are realized only through active and ongoing struggle over their creation, uptake, and revision. New technologies go through what sociologists of technology call a period of "interpretive flexibility" during which it is still not clear which social actors will have a role in stabilizing the new technology's meaning and form. As time goes on and different social groups work to stabilize and contest the technology, we move into a period of closure and stabilization. Trevor Pinch and Wiebe Bijker (1987) have described this process in the case of the bicycle in the late nineteenth century, which exhibited a wide range of design variation until it was stabilized into the low-wheeled form with air tires that we still see today. I consider this recognition of the socially embedded nature of technology as one of the core theoretical axioms of contemporary technology studies and as foundational to the theoretical approach I take in this book. In this approach, I draw from social studies of technology that see technology as growing out of existing social contexts as much as producing new ones (e.g., P. Edwards 1995; Hine 2000; Lessig 1999; Miller and Slater 2000).

It may seem self-evident that the representational content of media embodies a certain point of view and set of interests, but what is often less visible is how media is embedded in structures of everyday practice, particular technological forms, and institutional relations. We often see issues such as representations of women or violence in games taking the fore in social controversies, but in this book I argue that the broader institutional and business contexts of software production, distribution, and consumption are also important sites of contestation over the meaning of games. In order to build an agenda for how games can contribute to systemic change in learning and education, it is necessary but not sufficient to analyze representational content and play mechanics. In addition, we need to understand the conditions under which the game has been produced, advertised, and distributed, as well as how it gets taken up, domesticated, incorporated, and regulated in different contexts of play. In their textbook of cultural studies, Paul du Gay and his colleagues (1997) describe a "circuit of culture" that includes processes such as production and design, advertising, uptake, and regulation. They suggest that in order to understand the meaning of a new technological artifact, we must analyze all nodes in this circuit as well as the interaction between the nodes. In other words, we must realize that a game's design has a structuring but not a determining effect on how the game will be marketed or played, just as existing practices of gaming or education have a structuring but not a determining effect on what kinds of games will get created. In the case of commercial media, although the representational content such as characters and narrative are constantly in flux, the industry relations, distribution infrastructure, patterns of player-viewer engagement, and genres of representation tend to be conservative and deeply engrained within existing social and cultural structures.

The challenge in conducting this kind of analysis is to take seriously the powerful position occupied by the media text and industry apparatus, while also accounting for the role of everyday activity and the agency of media consumers. Within media studies, scholars have documented the ways in which audiences interpret and reposition cultural meanings in ways that the media text only partially determines (see, e.g., Buckingham 1993; Morley 1992; Radway 1991). Granting agency to the audience is particularly significant in the case of children, who have often been defined as passive and innocent receivers of media messages (see Jenkins 1998; Kinder 1999) and as "learners" and "becomings" rather than full social beings who actively construct social and cultural worlds (see James, Jenks, and Prout 1998). In their influential book *Situated Learning*, Jean Lave and Etienne Wenger (1991) argue that learning is an act of participation in communities of practice rather than an individual, cognitive process of internalizing knowledge. This focus on social and cultural context rather than on individual internalization has much in common with work in media studies that argues for the audience's more active and constitutive role. Henry Jenkins (1992, 2006) has put forth the notion of "participatory media cultures," which he originally used to describe fan communities in the 1970s and 1980s and has recently revisited in relation to current trends in convergence culture. Jenkins traces how fan practices established in the TV-dominated era have become increasingly mainstream due to the convergence between traditional and digital media. Fans not only consume professionally produced media, but also produce their own meanings and media products, continuing to disrupt the culturally dominant distinctions between production and consumption. A notion of "participation," as an alternative to "consumption," has the advantage of not assuming that the child is a passive observer or a mere "audience" to media content. It is agnostic as to the mode of engagement and does not invoke one end of a binary between structure and agency, text and audience. It forces attention on the more ethnographic and practice-based dimensions of media engagement and queries the broader social and cultural contexts in which these activities are conducted. A notion of participation is also well suited to the specific characteristics of digital, interactive media. In the consideration of media that require players' activity and are reconfigurable in various ways, models that rely on "reception" and "audiences" fail to capture the roles that game players and technology users occupy (Buckingham and Sefton-Green 2004; Ito 2008).

In addition to arguing for a participatory frame in understanding new media engagement, this book also makes an effort to link this participation to institutional structures and resilient patterns in our culture that contextualize the specific media texts in question. A growing body of ethnographic work in media studies has gone beyond an examination of the audience-text relation to consider the relation between audience and cross-cutting discourses and institutions. For example, Purnima Mankekar (1999) relates audience interpretation of an Indian soap opera to discourses of nationalism and gender, and Lila Abu-Lughod (1995) analyzes Egyptian television in light of contestations over modernity, femininity, and relations with the West. This work is part of an effort to take seriously the role of media in our everyday lives, while also accounting for the highly distributed and often contingent nature of its effects. S. Elizabeth Bird writes, "We really cannot isolate the role of the media in culture, because the media are firmly anchored into the web of culture, although articulated by individuals in different ways. We cannot say that the 'audience' for Superman will behave in a particular way because of the 'effect' of a particular message; we cannot know who will use Superman as some kind of personal reference point or how that will take place" (2003, 3). Other scholars have looked not only at the cultural web of meanings in which media is entangled, but also at the institutional contexts that structure media. Theories about how media and technology become domesticated in the home or appropriated by particular social groups are an example of this kind of social analysis. Sonia Livingstone's (2002) study of how families adopt new media and Stewart M. Hoover, Lynn Shofield Clark, and Diane F. Alter's (2004) analysis of the intersection of parenting discourses and media regulation in the home are examples of work that has taken strides in this direction. Roger Silverstone, Eric Hirsch, and David Morley argue that the technology "has the capacity to reproduce the social and political values of the society that created them." At the same time, they argue that this " 'politics of technology' . . . is not a totalitarian politics. It has to be understood in its full range, as a mixture of strategy and tactics, subject to the passage of time, and vulnerable to the capacity of local and domestic cultures to spoil or redefine the political and cultural inscriptions which we can now begin to decipher in the structures of the machine" (1992, 3).

Scholars in both technology studies and media studies have been arguing for the importance of analyzing in a grounded way the broader web of social and cultural connections that determine the shape and meaning of new media and technologies. Rarely, however, has this kind of analysis been carried out in a way that links detailed ethnographic fieldwork of media engagement with analysis of institutions linked but external to the site of reception, such as media development, marketing, and distribution. This book is an effort to address this conceptual and methodological problem by conducting an unconventional kind of genre analysis that cuts across various points in the circuit of culture. I use the concepts "media genres" and "participation genres" to read across different social boundaries and to describe how culture gets embodied and "hardened" into certain conventionalized styles of representation, practice, and institutional structure that become difficult to dislodge. I draw from John Seely Brown and Paul Duguid's (1996) notion of genre as something that cross-cuts form and content in media artifacts. For example, in the printing of a book, genres involve things such as typography, layout, paper weight, and binding, "the peripheral clues that crucially shape understanding and use" (132). Participation genres similarly involve the explicit content or focus of an activity as well as the subtle stylistic cues such as stance, gaze, and attitude that help us recognize a specific action as part of a category of practices. In addition to being a property of a media artifact, genre is also a mechanism for linking the artifact to a marketing and distribution network. Buckingham describes genres as "a form of contract with the industry and the audience: like the star system, it serves an economic function for the industry as a means of regulating the market, while also ensuring that audiences know (more or less) what kind of pleasure they will be getting for their money" (1993, 137). In this sense, participation genres do work similar to that of concepts such as habitus (Bourdieu 1972) or structuration (Giddens 1986), linking specific activity to broader social and cultural structure. More closely allied with humanistic analysis, however, a notion of genre foregrounds the interpretive dimensions of human orderliness. How we identify with, orient to, and engage with media and the imagination requires acts of reading and interpretation. We recognize certain patterns of representation (media genres) and in turn engage with them in routinized ways (participation genres).

If we look more carefully at the arguments for and against the transformative potential of new media in learning, this genre framework helps untangle some of the seemingly contradictory findings and incommensurable viewpoints among them. Although a focus on genre and structure may seem to privilege conservative tendencies, my goal is not to debunk the innovative potential of new media. Productive debate over the role of new media in transforming learning and education is often stymied by a lack of clarity over what kinds of social, cultural, and individual learning outcomes are up for grabs. Boosters of new media will point to local learning outcomes or innovation in technology or educational program design. Debunkers will point to the conservatism of entrenched institutions such as the school and capitalist relations. Both of these dynamics are at play in the story I tell about children's software, and both represent part of the picture of how a new set of technologies was incorporated into our everyday lives. To end here, however, would be to retell a familiar story about the struggle between structure and agency, between entrenched institutions and local innovation. Instead, I propose a different lens for viewing this story.

The struggle I describe is not about active audiences, participatory players, and innovative designers pushing back at the conservative structures of established industries and the expectations of educational institutions. Rather, it is about how *particular* kids and educators at local sites of media engagement and *particular* actors within technology development and media marketing become enlisted in shared sociocultural formations or genres. Within education, the debates among behaviorist, constructivist, and sociocultural forms of learning theory are cut from the same social and cultural cloth as efforts within the software industry to define the academic, entertainment, and construction genres and as the struggles within families to regulate the balance among schoolwork, play, and self-expression. The key battles that this book describes are not between the audience and the text or between the player and the industry, but between different genres that span production, distribution, marketing, consumption, appropriation, and domestication of software. Both change and the maintenance of stability are outcomes of successful enlistment of a distributed network that is mediated by new media, represented in compelling cultural discourses, embedded in powerful institutions, and performed at local sites of activity. Local innovation or the influx of new technology has no systemic effect unless it links up with other sympathetic parties and practices across the circuit of culture. In the remainder of this introductory chapter, I describe the multisited fieldwork that I conducted in order to investigate this distributed network of relations. The rest of the book discusses these networks as a set of distinct social, cultural, and technical genres.

The Study

This work draws from an interdisciplinary methodological frame that weds ethnography with approaches in technology and media studies that trace highly distributed and technological mediated forms of culture and practice. Anthropology has been grappling with methods for studying culture as it is distributed across wide-ranging networks of media, migration, and commodity capitalism. Calls for multisited ethnography (Marcus 1995; Martin 1994) and anthropological attention to objects such as the state (Gupta 1995; Gupta and Ferguson 1992), commodities (Appadurai 1986; Miller 1997), mediascapes and technoscapes (Appadurai 1990; Appadurai and Breckenridge 1988; Fischer 1991), online communication (Escobar 1994; Marcus 1996; Miller and Slater 2000), and large institutions (Nader [1969] 1972) have stressed the importance of reshaping anthropology to address ethnographic objects that are multisited and technologically mediated. The work that I draw on in media and technology studies rely on ethnographic approaches that are in line with this mode of inquiry. This book's ethnographic innovation is not in describing the unfamiliar, but in tracing unconventional linkages between familiar but dispersed objects and in describing how acts of play and consumption are related to the structure of technology development and distribution. In contrast to contextualization in a geographic area, George Marcus has argued that "within a multi-sited research imaginary, tracing and describing the connections and relationships among sites previously thought to be incommensurate *is* ethnography's way of making arguments and providing its own contexts of significance" (1998, 14).

The relationships I trace between everyday contexts of play, marketing materials, and the activity of software development are an effort to bring to the surface those connections across contexts that may not be immediately obvious. This effort involves tracing linkages across multiple local sites rather than analyzing the relation between local action and global structure. It resists the notion that a systemic structure exists independently of these local linkages, instead arguing that the sedimentation of local action itself is what constitutes and maintains broader structural patterns in society and culture.

More concretely, this method of fieldwork has meant engaging in a project that dips into multiple field sites that are spatially distributed and separated by institutional boundaries, but are also part of the circuit of culture that constitutes children's software. These sites include after-school clubs where kids and adults engage with children's software, software design and development companies, and marketing and distribution channels for these products. Although my fieldwork in these different sites was not symmetrical—I spent much more time on the player side—I have tried to capture each site's local contingencies. The greater attention to the player-side fieldwork has to do with the conditions under which I conducted the fieldwork as well as with my commitment to capturing the agency and voices of children and to giving them a seat at the table in the debates over new media and learning. Creators and marketers of software and educational reformers have existing platforms through which to disseminate their views. One important role of ethnography is to make visible everyday activity and differing perspectives that are otherwise inaccessible.

Work in the "new paradigm" of the sociology of childhood (James and Prout 1997) has noted the ways in which children's voices and agency are often erased in research. Rather than considering them as complete "beings" who live full social lives rich in culture and knowledge, most work on children considers them incomplete "becomings" on their way to full social and cultural participation (Cosaro 1997). Studies of children are dominated by psychological approaches that look at childhood in terms of "ages and stages," and that do not take into account the ways in which children are productive actors in the social world (James, Jenks, and Prout 1998). My work documenting the ways in which kids produce meaning through play, appropriating and reshaping the meanings of technology in local micropolitics, is an effort to consider the significance of kids' agency in the here and now of their everyday lives. In line with this approach, I often use the term *kid* rather than *child* because *kid* is the term that school-age children tend to identify with themselves (Thorne 1993, 9). I turn now to a description of the field sites in which I encountered kids and their engagements with children's software, and thereafter to a description of the fieldwork I conducted on the industry side.

The Fifth Dimension
The Fifth Dimension (5thD) consortium is an influential educational reform project with a history spanning decades. Michael Cole, who runs one of the sites that I studied in southern California, is the central organizing figure in this effort, although many other researchers have been involved in the project through the years (Cole 1994, 1997; Cole and Distributed Literacy Consortium 2006; Engestrom 1993; Nicolopolou and Cole 1992; Vasquez, Pease-Alvarez, and Shannon 1994). Since the late 1970s, the 5thD has evolved from a single experimental after-school cooking club for kids to an international educational reform effort that makes use of computers and telecommunications. Club sites are located across the United States as well as overseas. They are funded through grants by foundations such as Mellon and Russell Sage, the government, universities, and a variety of efforts by local communities. The university professors, graduate students, and other staff who work with the 5thD project hold joint identities as local implementers in close touch with the day-to-day operations of the clubs and with the communities in which they are located and as researchers with a reformist voice in the educational research community. The 5thD clubs have become an increasingly influential reform effort through the strong backing by the educational research community and by the local communities in which the clubs are located.

From the outset, the 5thD effort has combined theory and practice. The 5thD is based on a strain of psychological theory informed by the works of Lev Vygotsky (1987) and by the school of Soviet psychology with which he is associated. In this view, learning is best understood and supported as a socially and culturally situated act of engaging with other people and artifacts rather than an act of individual knowledge acquisition. The 5thD philosophy draws in particular on Vygotsky's theory of learning through the "zoped," or the "'zone of proximal development': the place at which a child's empirically rich but disorganized spontaneous concepts 'meet' the systematicity and logic of adult reasoning" (Kozulin 1986, xxxv). The 5thD activity system has been designed to enable zopeds where children and adults can engage with a computer in joint tasks. The undergraduates staffing the sites are concurrently taking a course in cultural psychology where they are exposed to sociocultural approaches to learning, and their work in the 5thD is framed as observational fieldwork. The 5thD is a laboratory for testing the theories of cultural psychology as well as a way of realizing them in practice.

"Cultural psychology" in the vein of Cole and his collaborators (Cole 1997) found a common approach in anthropological approaches to learning and education, and has been a rich contact point between the disciplines of anthropology and developmental psychology (Brown, Collins, and Duguid 1989; Cole, Engestrom, and Vasquez 1997; Lave 1988, 1993; Lave and Wenger 1991). This interdisciplinary intersection of "sociocultural learning theory" takes a reformist stance toward mainstream American education's psychologization and individuation of education and assessment, arguing for the importance of social and cultural factors in learning (Lave 1988; Lave and Wenger 1991; Varenne and McDermott 1998). Ethnography becomes a method for observing "cognition in the wild" (Hutchins 1995) and for accounting for factors that are systematically excluded from experimental psychology (McDermott, Gospodinoff, and Aron 1978). This line of research also challenges the Piagetian tradition dominant among many educators. Whereas Jean Piaget sees the engine of learning in a child's organic, individualized, but universal developmental readiness, sociocultural learning theory sees learning and development as a social act that varies across different cultural contexts.

I conducted research on the 5thD as part of an evaluation project that sought to document learning processes at the sites through video-based analysis. This project was collaborative, run out of the Institute for Research on Learning (IRL) together with the Laboratory of Comparative Human Cognition at the University of California, San Diego. The bulk of the field-

Figure 1.1
Picture-in-picture format for viewing videotape. SimCity 2000™ © 1993–1994
Electronic Arts Inc. *SimCity 2000* and *SimCity* are trademarks or registered trademarks
of Electronic Arts Inc. in the United States and/or other countries. All rights reserved.

work I report on was conducted in the 1995–1996 academic year. Together
with a team of researchers, I was involved in videotaping interactions
around computers at several 5thD clubs. We taped the social interaction
around the screen and captured a scan of the screen itself, then we merged
these two tapes into a picture-in-picture format for our review and analysis
(figure 1.1). Our team also had access to the on-site field notes that the
undergraduate participants wrote up every day that they participated at
the site.

The site that was most central to this research and also the oldest current
5thD site is located at a Boys and Girls Club off a busy thoroughfare in a
suburban neighborhood in Southern California. Between three and four
o'clock, when I would usually arrive at the club, kids are walking in from
the school across the street or are being driven in by parents or other
caretakers to spend their hours between the end of school and the end of
the parental workday. The Boys and Girls Club is a "safe" place for kids to
spend this two-hour or more gap in their day. If one enters the club during
these hours, one is greeted by a cacophony of kids' voices as they mill
about the club's central room, which is occupied by pool tables, a few
coin-op arcade games, fooseball, soda and candy machines, and clusters of
sofas. Adult staff circulate among the kids, playing with them, organizing
activities, and occasionally acting as disciplinarians.

At the far end of the main room is a door to the art room, filled with art supplies of various sorts as well as computer parts from old Apple computers and PCs that kids are allowed to tinker with during special workshops. Outside, on the other side of the building, are a basketball court and a pool. Off to one side is the door to the library and the club's "educational" area, which includes the room where the 5thD site is housed. Half of this educational area is occupied by bookshelves with reference books and by round tables generally devoted to homework. On this side is also a small computer area that is not part of the 5thD club, with a PC connected to the Internet and some Mac Pluses for doing writing assignments. On the other side of the room, separated by a cubicle-like partition, is the 5thD. Tables rim the 5thD space, occupied by ten or more computers and a printer or two. One table off to the side also houses the folders for each 5thD citizen, which track their progress through the games and other activities. It also houses "the maze," a wooden structure with miniature rooms and doors where kids move small objects called "cruddy creatures" that represent the kids themselves. As they progress in the club's activity system and master different games, they move their cruddy creatures through different rooms of the maze. A poster above the maze describes the "consequences chart" that dictates what games they can play in each room and how to proceed from room to room. Playing with commercial children's software dominates the activities at the site, but there are also rooms in the maze that propose playing board games, making a video, and doing other activities not on the computers. When a kid completes all rooms of the maze, he or she wins the title "Young Wizard's Assistant" and is able to play the high-end games at the site and is responsible for teaching others.

When the 5thD is not in operation, a piece of poster board closes off this area of the club. As 3:30 approaches, the time when the 5thD opens, some of the kids at the Boys and Girls Club begin to move from the basketball court and pool tables and start to mill around the 5thD area, attentive to the university undergraduates who are beginning to trickle in. The site coordinator is turning on the machines, getting the maze organized, and checking in on the undergraduates as they arrive. When the site opens, kids start pairing off with undergraduates, grab their folders, and eventually settle at a computer, while the site coordinator directs traffic. Things are chaotic and lively as kids and adults jostle around the computers and try to figure out what software is available on what machine or who is working with whom. Although some 5thD sites have a formal sign-up sheet to manage who gets to participate each day, participation at this particular site is managed informally, and kids jostle for position at computers and with undergraduates. Some kids might also be writing e-mail to other 5thD sites or to the Wizard, the mythical entity who oversees the clubs and sometimes appears on a live computer-based chat. Some of the 5thD clubs do not allow free circulation between the 5thD and the Boys and Girls Club at large, but at this site both kids and undergraduates circulate between the 5thD, the library area, and other parts of the Boys and Girls Club. Kids often stop by briefly to observe or intervene in other kids' play or get bored at a particular game and decide to go play pool or to mingle with kids at other parts of the clubs. Kids might also decide that they would rather play a game with a friend, leaving some undergraduates with no kids to partner, so that they end up playing with each other, trying to learn a new game. There can be as few as three and as many as fifteen kids at the 5thD site. On a crowded day, latecomers have to wait for their turn with an undergraduate and often stand observing other kids play.

This particular club caters mostly to middle-class, white and some Latino kids whose working parents need an after-school activity for their kids. In contrast to the kids at some of the other 5thD sites, many of the children attending this club have computers at home, so the site has become an extension of home entertainment activities and an opportunity to engage in these activities in a peer group. A small but highly visible group of technically savvy older boys, ages ten to twelve, often dominate the higher-end machines at the club, exchanging tips and opinions about current games. Girls attend the club in somewhat lower numbers than boys. The dominance of the older boys in the group led for a brief period to the institution of a "girls only" day each week, where girls could play on the machines that they wanted without having to negotiate with the boys. Familiar gender dynamics are present in the male dominance of technology at the club, even though club operators attempt to increase girls' representation.

Less than a mile away, another site where I observed serves a predominantly working-class Latino community. At this site, the 5thD operates in a trailer that it shares with a Head Start program, located on the property of a church. The children who attend this 5thD site come for the 5thD activities alone rather than to use other facilities, as is the case at the Boys and Girls

Club. The site coordinator is the mother of one of the children and speaks Spanish and little English. During 5thD operations, a lively mix of languages can be heard as English-speaking university students mingle with bilingual children and the occasional Spanish-speaking parent or community member. All the curricular materials are available in both English and Spanish. The space is cramped, with computers rimming each wall around the round central table for activities such as drawing and board games. There is a stronger presence of parents and families at this site because of community ties through the church, and active efforts have been made to recruit parents to help staff the site. Older children sometimes bring their small siblings, and children just starting kindergarten and learning to read are often there as well. With the exception of a few Spanish-language games, most of the software is similar to the software at the Boys and Girls Club site, as are the curricular materials. In contrast to the Boys and Girls Club site, however, most of the children at the church site club do not have access to computers at home, and their parents generally lack access to them at their workplaces. The site became an opportunity for both kids and their parents to engage with new technology that they would otherwise have limited access to (Vasquez, Pease-Alvarez, and Shannon 1994).

The third site at which I observed was one that I, my advisor Ray McDermott, and other IRL staff (Anne Mathison, Shelley Goldman, Dena Hysell, Charla Baugh, Gary Geating, George Lopez, and Ingrid Seyer) started in Menlo Park. Our first session was in the fall of 1995, the second year of the project. This effort was motivated by my desire to interact with my research subjects on a more regular basis and by the spirit of community service. In alliance with the East Palo Alto Stanford Summer Academy (EPASSA), a summer camp run by Stanford University, we recruited middle-school kids from East Palo Alto and Redwood City to come to IRL on weekends and play with educational games and Internet technologies. The EPASSA staff helped us coordinate a meeting at which the EPASSA families could see what was happening in the club and we could explain our research.

Of the fifteen or so families that attended the informational session, six became regular participants and occasionally brought their friends. All participating families were middle-class and working-class African Americans, with the exception of one Latino family. Other than the Latino boy, the kids did not have access to computers at home and had limited exposure at school. Project members and a number of other people hired on a temporary basis staffed the club in lieu of the undergraduates who generally act as tutors at 5thD sites. Because we were also dealing with an older age group and had access to the high-end technology at IRL, we were able to experiment with more kinds of software and activities than other 5thD sites have, including Web page design and digital video production. Throughout the 1995–1996 school year, we met weekly, with between three to seven kids, and most sessions were taped. As the year progressed, many of the kids would also appear at IRL during the week to make use of computers for homework and for fun. During the summer of 1996, the club's core members ran their own summer program at IRL, working as teachers for kids enrolled in a special EPASSA elective. The kids from this local club are in many ways my primary informants, although they are not necessarily the ones most heavily represented in this book. They are the kids I not only observed through the video record, but also interacted with on an ongoing and regular basis as informants, interlocutors, and video-gaming companions.

The 5thD is a unique setting that enabled me to observe interactions with children's software in a social environment, but it differs in many ways from the place where more conventional interactions with computers take place—the home. There is a stronger peer-group influence in the 5thD than in the home, and the adult presence in the form of undergraduate tutors and educational researchers is unconventional. The 5thD provides more access to a greater variety of software products in the genres that I am investigating, whereas in the home, educational software more commonly plays second fiddle to console gaming and television. The 5thD magnifies children's interest in educational genres of software by creating a peer-status economy around play that would otherwise gravitate toward mainstream action games and television. The adult supervision also emphasizes the academic content embedded in the games, making the 5thD's overall setting more attentive to educational concerns than are most home contexts. Another difference is that some of the social distinctions that operate within unsupervised play contexts are erased. Girls are actively recruited to engage with the technology, and at many of the sites computer access is given to children who do not have computers at home. In contrast to schools, the 5thD provides a more fluid and open-ended environment where kids can move between activities and engage in social interactions with other kids and adults.

To borrow a term from Joseph Giacquinta, JoAnne Bauer, and Jane Levin (1993), the "social envelope" surrounding computing in the 5thD is distinctive. These authors' work documents how use of educational computing in the home is rare, in large part because of the difficulty of choosing and purchasing these products and because parents need to encourage kids to use such products even when they are present in the home. The 5thD, by contrast, is an environment that maximizes engagement with the media that was the focus of my fieldwork. Although this context was idiosyncratic, it did provide a rare opportunity to conduct extended observations of how kids engage in these products. The 5thD effort is part of the same cultural fabric and historical conditions that gave rise to the children's software industry. Both the organizers of the 5thD and the early developers of children's software were part of a movement toward more progressive and child-centered forms of education that made use of new technology as an ally in their efforts. In other words, the 5thD was not simply a "neutral" site in which to observe play, but was itself part of the broader set of institutional conditions and discourses that I analyze in this book.

Industry Research

The industry side of my research relied on a more limited ethnographic toolkit and was more targeted than the work that I conducted at the 5thD, so I can describe it more briefly. After completing fieldwork at the 5thD, I began an effort to document what happens on the other side of the circuit of production that brings the software to the kids' hands in the first place. I began research on the history and context of the software industry, conducted a series of interviews with software developers between 1998 and 2000, and observed and had a series of informal conversations with people at trade shows and industry events. I conducted most of the interviews in person in the greater San Francisco Bay Area, but did four phone interviews as well.

In addition to the interviews, I conducted a literature review on children's software and the industry. I looked at Web sites with industry statistics and news, as well as at bulletin boards that discussed related topics. In contrast to the mainstream entertainment industry of computer gaming, relatively few publications deal specifically with children's software, and much of the information I was able to find was on targeted Web sites oriented toward tech-savvy parents and teachers. I turned to news

media for access to much of the industry backstory that was otherwise
locked up in proprietary reports. I conducted a review of references to the
industry in the *New York Times* and *Wall Street Journal*. With the help of
two research assistants, I did a comprehensive search in these two publications for appearances of the topics "children's software" and "computer
games" in the period from the early 1980s to the year 2001; we also
searched for all of the key corporations I was tracking. Although there are
many gaps in the public record, I was able to gain an overview of the
corporate players involved. I also reviewed advertisements for children's
software over a period of a year and a half in 1999 and 2000 in the magazine *Family PC*.

Organization of the Book

The next three chapters draw from my research on both the 5thD and the
children's software industry, organized by the focal three genres: academics, entertainment, and construction. The academics genre, discussed in
chapter 2, is centered on software to support school-based content and is
marketed to middle-class families oriented to academic achievement. The
history of this genre traces the early origins of edutainment and how it
was transformed from an experimental educational reform movement to
a niche industry that now relies on established formulas developed in this
early period. In their play with the games in this genre, kids are quick to
recognize each game's achievement goals and to orient themselves to fulfilling the conditions to move ahead in a task and "beat" the game.

Chapter 3 describes the genre of family-friendly entertainment that
emerged in tandem with the development of more sophisticated multimedia capabilities in PCs. In contrast to the software titles in the academic
genre, the titles in this genre rely more on open-ended exploration and
do not focus on delivery of school-based content. They are closer to
the family-friendly entertainment characteristic of children's television.
Educators and developers of this genre of software appeal to parents'
desires to support children's play and pleasure. Although adults will try to
orient kids toward the more school-like content in some of these games,
kids tend to see play in this genre as a process of open-ended exploration
focusing on the pleasure in visual and interactional special effects.

Chapter 4 examines the construction genre, which has its roots in educational efforts to promote computer programming. Games in this genre involve tools and simulations that enable kids to author, construct, and manipulate digital media; they appeal to identities of technical mastery and empowerment and are marketed based on their ability to provide tools for creativity and self-expression. Kids' engagement in this genre often involves projecting their own lifeworlds and identities into the online space, in often unpredictable ways.

The conclusion returns to some of the issues raised in this introductory chapter, examining the ways in which this case study of children's software can inform unfolding developments in the area of new media and learning.

2 Academics

In an article for *Byte* magazine published in 1984, Ann Piestrup describes a new media category that she calls "graphics-based learning software." "Only recently are computer scientists and educators beginning to collaborate to create learning software that can fulfill the promise of the personal computer to transform education." She argues that unlike text-based computer-aided instruction approaches or entertainment titles that require little interaction on the part of the child, "powerful learning software programs, such as learning game sets and builders, use graphics to convey meaning, not to decorate the screen" (1984, 215). Her article reviews software titles produced at the company she founded in 1979, renamed The Learning Company (TLC) in 1983.

I met with Piestrup, who now goes by the name McCormick, in 2000, at Buck's Café in Woodside, California, at the peak of the dot-com deal making. She reflected on her experience in the 1980s and described the heady sense of excitement felt at the time in creating a new category of media and a new category of learning experience that differed, on one hand, from instructional software being used in schools and, on the other, hand, from video games. "We created a new category by working with an Atari game designer and educators that were serious. We weren't trying to mimic zooming video games, but we were mimicking real-time interactivity." Moreover, she said, "I didn't want to call it educational because to me that meant schooling, dusty, institutional. That's why I called [my company] The Learning Company not the Education Company." TLC went on to become one of the largest names in children's software and was sold to Mattel in 1998 for $3.8 billion. Although the company has weathered many ups and downs and has been bought and sold numerous times since then, it is still one of the leading brands in children's software.

Despite being burned numerous times in business dealings through the years, McCormick was in 2000 still an impassioned entrepreneur and spokesperson for the uses of computers to support learning. She showed me her new business proposal to create new learning environments that make use of the growing power of today's PCs and networking infrastructures. "I want every child in the world to be able to get the basic skills they need to function thoughtfully, with graceful feelings as well." A former nun, schoolteacher, and educational researcher, McCormick is an irrepressible missionary for the cause of computers in enhancing learning, particularly for disenfranchised populations: "We want to do lifelong learning for the whole world. And there is assessment going on constantly, and we make sure they [children] move all the way through the math, science, and readiness that they need. We think about the beauty of the structure . . . and not just the nuts and bolts." She sees her work as a quest for human equality, based on "a conviction that stems from my sense of human fairness that extends to all children." For her, literacy is a basic human right that ensures a voice in the social world.

McCormick embodies the passion and dedication of the early developers of learning software who felt that computers could enable child-centered, egalitarian, and engaged approaches to learning. Her challenges in realizing this vision in the commercial sector also point to the contradictions and tensions inherent in crossing the boundaries between school and home, education and entertainment, nonprofit and for-profit realms. This chapter explores the negotiations among educators such as McCormick, capitalist enterprise, and the changing structures of technology in the emergence of a category of software that came to be called *edutainment*. I use the historical backdrop of edutainment as a way of exploring the academic genre of children's software. Unlike the entertainment and construction genres, which I describe in chapters 3 and 4, the market category "edutainment" and the academic genre are firmly grounded in an educational imperative to instruct children in literacy skills and core academic content. Beginning with a discussion of historical roots, this chapter describes the cultural context and social distinctions related to this genre of software and how they manifested in everyday play in the 5thD during the period that I did my fieldwork.

The Historical Roots of Edutainment

Learning software for children is contextualized by discourses of childhood, learning, and play that framed earlier media such as children's literature and developmental toys. As with these earlier forms of children's culture, children's software was initially conceived of as an educational tool for children that wedded the virtues of play, learning, and literacy, and that drew from a growing twentieth-century belief within the middle-class American home that learning should be fun to be effective. Freed from the classroom's narrow curricular constraints that defined early drill-and-practice computer-aided instruction, the new commercial software was designed to be appealing and engaging for children and to compete with other leisure-time activities. This dynamic negotiation among schools' educational demands, parents' achievement concerns, and children's desires and pleasure has been a central one in the lives of U.S. children at least since the late nineteenth century, defining the ways in which children's media and toys have been produced and consumed. McCormick and other early innovators in children's software occupied a shape-shifting patch of turf in this contested terrain, where children's media was designed not only to be entertaining and engaging for children, but also to appeal to parental concerns about learning and achievement.

Educational children's products have sustained themselves in a variety of forms as a niche market for educationally minded families through the years. Before the advent of children's software, there was both a well-established discourse of educative play and a related market in children's media and toys. Although edutainment represented a new set of technologies and a new market niche, its success was highly contingent on the fact that earlier media such as public children's television and children's literature had established certain genres that families and educators recognized as both fun and educational. After the initial experimental period, the children's software industry also utilized distribution channels such as bookstores and toy stores that were already in play for reaching families who were likely to purchase educational media (Buckingham and Scanlon 2003).

In book publishing of the seventeenth and eighteenth centuries, children's content was initially characterized by highly didactic, moralistic, and religious tracts. The mid–nineteenth century was a turning point in the production of media commodities directed at children, seeing the growth of fiction that was written to delight and engage children. Titles such as *Alice's Adventures in Wonderland* and *Fern's Hollow* were indicative of a freeing of children's literature from its religious and didactic roots toward a more playful and imaginative model. The next century saw a blossoming of children's fiction, the establishment of a related segment of the publishing industry, and the growth of a new genre of children's literature in the form of the comic book (Kline 1993, 89–97). In contrast to television and radio, books are considered a vehicle for achieving both basic and cultural literacy and thus have always been a preferred form of media for bourgeois sensibilities. Even without overtly didactic content, children's literature has occupied the privileged terrain of learning media, marketable as a highbrow commodity to educated families. Describing the current state of children's literature, Stephen Kline writes that the children's book industry in Canada and the United States is "a niche market, based on a narrow segment of the population buying a lot of books: mainly the wealthy and educated book-oriented segment of the market, people who still see books as vital tools of socialization" (1993, 96).

Toy consumers, by contrast, have a more mixed demographic, and learning toys are only a small—albeit resilient—segment within the broader toy industry. Gary Cross describes the growth of an American toy industry infused by mass media in the 1930s with the advent of Mickey Mouse and Shirley Temple dolls. He also describes an alternative trend in toy production, however: "Not all parents in the 1930s bought their children Mickey Mouse hand cars and Shirley Temple dolls. While toymakers were selling Brownies and Kewpies, psychologists and teachers were promoting plain wooden blocks and pegboards as early learning tools" (1997, 121). According to Cross, significant growth occurred in the number of new parenting experts and manuals in the 1900s and in a rational approach to child rearing that he calls "scientific motherhood." Contemporary efforts to make learning enjoyable can be placed within an established educational tradition that includes Jean Piaget and Friedrich Froebel, who believed in the educational potential of play. As children were removed from the workforce, parents increasingly saw play as the core activity of childhood (Cross 1997, 123–4). Ellen Seiter analyzes advertisements in *Parents* magazine during this period of growth in the number of educational toys from the 1920s to the 1950s. "*Parents* continually repeated the platitude that play was educationally valuable." "Toys could guarantee joy yet be instruments of hard work and achievement. What more could anyone ask from a commodity?" (1995, 66, 67). As Cross puts it, "Play had become the 'work' of children. And work required tools" (1997, 129). He describes how parents with more Victorian values were faced with the task of resisting a rising tide of children's consumer culture:

To many middle-class parents that consumer culture seemed to express the narcissism and quest for immediate gratification that bourgeois Americans identified with the lower class. And it threatened to engulf their children as they went to the movies and ached for those flashy toys offered by Louis Marx. The ideals of self-directed play, with objects of simple design had nothing to do with the appeal of character toys. Educational playthings represented, to middle-class parents, a bulwark against the tide of commercialism and its threat to undermine parental authority and Victorian values. (1997, 134–135)

This bourgeois view of childhood play as a privileged and generative site for developing the agency of cultural producer or "worker" was established in opposition to a hedonistic, "consumptive," or "recreational" view of play that was associated with licensed products and children's "junk culture." This period saw the emergence of the contemporary cultural distinction between high and low children's culture and the integration of this distinction with processes of class distinction.

After the ascendancy of television in the 1950s, these cultural and social dynamics changed quite dramatically, and the Victorian parental orientation toward childhood discipline and intellectual development was overshadowed by the influence of a fast-paced, commercial, fantasy-based children's popular culture. Attitudes toward restraint and denial in children's consumption eroded in the face of television and the ubiquity of children's popular culture. Educational toys were marginalized in an era of novelty toys and discount toy retailers, though they were still an important niche market, particularly for preschoolers. As commercial children's culture has taken hold, however, many families have been part of a countervailing tide of resistance to commercial children's culture. A large volume of publications aimed at the educated middle-class argues against children's exposure to media and licensed commodities, ranging from conservative calls for a return to family values to left-wing attacks on negative stereotypes in commercial media. The market niche of educational children's products, ranging from wooden train sets to classic children's books and Lego blocks is a source of cultural capital that unites the anti-commercial sentiments of certain sectors of both the conservative and the progressive middle class (Seiter 1995, 3–6).

In her study of the relationship between class identity and parenting, Annette Lareau (2003) describes how middle-class families engage in a process of "concerted cultivation," where parents manage and structure their children's time outside of school. Structured activities such as music lessons and organized sports are hallmarks of how middle-class families manage and "enrich" their children's lives. She contrasts these families to working-class families, in which parents see themselves as facilitators of "natural growth" and kids have more unstructured time to spend with friends in a way that is not managed and directed by adults. These differing orientations to child rearing are also reflecting in attitudes toward media, where middle-class families tend to set limits on unstructured recreation and screen time. Seiter has critiqued the stance of members of the educated middle-class who feel that management of their children's media environments is an indicator of superior parenting: "It is necessary continually to attack the smug self-satisfaction of educated middle-class people who believe themselves to be cleverer than those who do not attempt to monitor, mask, or deny their own television viewing, who believe that other people's children are already ruined by 'exposure' to television" (1995, 6).

Lareau and Seiter's work provides an important reminder that children's media consumption is inseparable from specific parenting approaches and that these approaches are in turn deeply implicated in the production of class identity. The development and marketing of children's software is likewise inseparable from these class dynamics and parenting attitudes and from a much longer cultural history of how middle-class families have valued educative forms of play. The dynamics are also clearly evident in the public scripts about regulating media in the home that were the subject of Hoover, Clark, and Alter's (2004) study. In these scripts, many parents referenced the notion that limiting media is a mark of good parenting, even while the actual practices in their households rarely conformed to these public scripts. Sonia Livingstone's study of families and new media adoption similarly indicates that the relationship between class and media consumption is not straightforward, but that middle-class parents tend to mobilize scripts that have to do with limiting media use, working to keep their children occupied with more "productive" activities (2002, 93–99).

Digital media entered into these existing social and cultural distinctions by introducing flashy new forms of entertainment media—video games and computer-based media—that held out educational promise. Creators of educational software incorporated the orientation to play and many of the visual elements of video games, but framed their products as being educational and enriching. Like the educational toys of the nineteenth century, educational software was and still is seen as a bulwark against video games and repetitive, hedonistic, and violent play. Software produced by companies such as TLC are played on computers rather than with game consoles, on the "good screens" in contrast to the "bad screens" of television (Seiter 1999, 247). Although mainstream commercial licenses are increasingly dominating children's software, companies such as TLC have tended to shy away from the commercialism implied in mass licensing, instead creating their own characters or linking up with Public Broadcasting System (PBS) content such as *Blue's Clues* and *Arthur.* They have worked to develop a genre of software that is fun and family friendly, but oriented toward concerted cultivation.

The ways in which early developers of children's software grappled with defining new media genres is a lesson in the domestication of technology—how existing social groups struggle to appropriate and position the meaning and value of new forms of media. These developers redefined video games into genres that were closer to educational and highbrow media, but they also appealed to childhood play in order to create a consumer product for home use. Even before the advent of children's software, the term *literacy* was attached to the use of computers, a cultural marking that differentiated it as a highbrow and "difficult" media form, structurally set off from "illiterate" and developmentally regressive forms of media such as television and video games. Gaining technical literacy through computer use is often associated with gaining traditional literacy through books. Children's software, packaged to be played on computers rather than on game consoles, incorporated these educational valences as well as content that was explicitly tied to traditional literacy and academics.

For early developers such as McCormick, the goal was to put technical tools in the hands of the disenfranchised and to alleviate the oppressiveness of dominant notions of education. Efforts toward technological empowerment can cut both ways, however, particularly when they are contingent on expensive consumer products such as computers and software that are available only to the wealthy. Unless reform efforts address marketing and distribution issues, technical literacy simply becomes one more element of cultural and material capital that reproduces class differences. As Hervé Varenne and Ray McDermott (1998) have argued in their description of "successful failure," constructions of intelligence and learning inevitably invite constructions of failure and social distinction, often thwarting educators' and reformers' best intentions. On occasions when new educational technologies have entered the scene, reformers have hoped to transform some of the underlying social dynamics that have alienated some children from academic achievement. More often than not, however, these efforts have tended to favor already privileged children (Seiter 2005, 2007; Warschauer 2003). These existing conditions structuring parenting, cultural capital, and children's play set the stage for the emergence of the new media genre of edutainment.

From Education to Learning and Back Again

At the same time that McCormick was producing software titles such as *Gertrude's Puzzles, Rocky's Boots,* and *Reader Rabbit,* other educational researchers at the University of Minnesota were beginning to commercialize products such as *Oregon Trail* and *Number Munchers.* The Minnesota Educational Computing Corporation (MECC) was originally funded by the State of Minnesota in 1973 and became a public corporation in 1985, riding the successes of these software titles. Jan Davidson, a former teacher, started her company Davidson & Associates in 1983, developing titles such as *Math Blaster,* which in its various incarnations has been the best-selling piece of math software through the years. These software titles, all originally produced for the Apple II, became the pioneers in the new market for educational software for home use. Although growing out of school-based uses of computers, these new products were designed for the home user and the consumer market. They departed from the strictly curricular and instructional goals of the majority of school-based software by incorporating visual and narrative elements from popular culture. For example, *Math Blaster* took a standard drill-and-practice instructional mechanism but embedded it in a shooter-game idiom.

The late 1970s and the 1980s saw the founding of experimental efforts such as the 5thD, Apple Classrooms of Tomorrow, the Vivarium project at the Open School in Los Angeles, and programs at the Bank Street College of Education, which piloted these new technologies in experimental educational settings. User communities and development communities were in close contact during the Apple II and early multimedia era of educational software. Bank Street developed its own software and operated an alternative school. Alan Kay, one of the developers of the Macintosh, participated in launching an educational technology program at the Open School in Los Angeles (Kay 1991). Seymour Papert, who developed the LOGO programming language at MIT, also ran educational programs in various schools with his technology (Papert 1980). Apple had a large education division that worked with the Apple Classrooms schools in developing curriculum and providing computers. And its research divisions were incubating the multimedia products that were to become the next wave of learning software. Although consumer products were being developed at this time, they were oriented to a small market of like-minded educators and parents. Because of Apple II's minimalist platform, development costs were low enough that extensive sales were not required to support development. Graphics were simple, but still managed to convey basic educational principles such as the logic of circuitry in *Rocky's Boots* (figure 2.1).

Most of the early innovators in educational software had backgrounds in formal education before turning to commercial efforts. These early

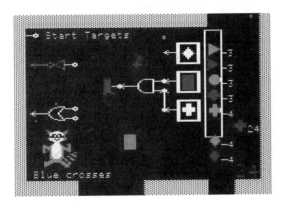

Figure 2.1
Screen shot from *Rocky's Boots*. Reproduced with permission from The Learning Company, Inc.

groundbreaking years in the industry were characterized by a sense of optimism and social mission. Ann Piestrup (McCormick), quoted in a Harvard Business School case study, described this sense of mission, tying together the heady promise of personal computing that was budding in the early 1980s with the educational mission of promoting active, engaged, and entertaining learning for children:

Our core values here involve our desire to prepare children for the computer age. We want to do that with technical excellence in computing. We want to use the very best mass market micros to do that, to do it playfully, engage the kids, involve them, get them excited about learning, give them an active goal so it's not a sugar-coated pill where there's some dinky reward or something. It's really involving children in a way that they become totally excited about learning and forget that it's a task. Using TLC programs is like building something with an erector set where you get totally lost in the process. So our goal is to offer that kind of learning on the computer specifically for skills that are needed in the future. No one is quite doing that, building thinking skills, ability to analyze, to construct, to approach things from different angles, to think flexibly, to reason carefully, and to do that in a way that you're building something, not destroying it. A real explicit value is: we don't accept software that blows things up. We don't like blowing things up because they are aliens. We like finding out about aliens! There's a lot of belief about our work being good for people and that really drives us. It isn't just selling soap. (Unpublished case study, 1984)

Jan Davidson, in an interview with *Children's Software Review*, echoed a similar sense of mission that was primarily educational rather than business oriented: "When we started the company [Davidson & Associates], I remember first having to make a big decision—'Am I going to be a teacher or a business person?' That was very hard. . . . I always thought of myself as a teacher and felt that I was betraying my goals by leaving the profession" ("A Conversation" 1997, 25).

Titles such as *Math Blaster, Reader Rabbit,* and *Oregon Trail* are considered classics and are still on the market today. Many of the products created for the Apple II are still considered among the best children's software; they have been upgraded and updated with newer graphics and sounds, but they still retain the same content and play dynamics. Elizabeth Russell, who was at TLC when I interviewed her in 1998, described her view of these "evergreen" titles: "One of the evergreen products here is the *Oregon Trail*. It's one of the oldest pieces of educational software, and it's still one of the best. It's twenty-six or twenty-seven years old, and teachers will still talk about this as the ideal of what a good piece of software is because kids apply math skills and thinking skills to real-world problems. And then they face the consequences. Those kinds of things make a program good."

In addition to producing classics such as *Oregon Trail*, this period of innovation saw the establishment of the basic formulas and genres of children's software that continue to be reproduced and repackaged today in a variety of titles. *Oregon Trail* originated the genre of educational travel adventure, where kids have to calculate their rations and supplies as they travel through a simulation of a historical journey. Similar titles such as *Amazon Trail* have followed in *Oregon Trail's* footsteps. *Math Blaster* represents a more behaviorist but entertaining drill-and-practice model where kids are given rewards for completing math problems: bullets that they can use to play shooting games. It is a more literal hybridization of the educational (drill-and-practice) and entertainment (shooting-game) idioms.

Although products such as *Oregon Trail* continue to be popular, the learning-software industry currently sustains itself on adventure games. Games in the *Math Blaster, Jump Start,* and *Reader Rabbit* series embed academic problems and tasks within an adventure-game format. This strand of software development has increasingly come to focus on curricular content rather than on open-ended game play and defines what I call the academic genre. The most typical design relies on academic minigames embedded in a role-playing scenario. Along the way to completing some kind of mission, the player encounters various problems or puzzles related to academic subject matter. These puzzles may be math problems, science questions, or reading games, but in general their content is unrelated to the role-playing fantasy narrative. Although these games do not take a narrow drill-and-practice approach, the educational philosophy behind them might broadly be associated with a behaviorist approach, where children are given external rewards (action games, eye candy, points, etc.) for completion of academic tasks. These games also generally keep close track of scores that are usually tabulated in a passport or report card format.

Although the earliest versions of these kinds of games were not produced to correspond to specific subjects for particular grades (fourth-grade math, second-grade reading, and so on), the later versions were, and the packaging features checklists of particular topics. The genre standardized around this form of game design and streamlined development around a successful formula. Companies committed to an underlying game engine into which different forms of academic content and minigames could be plugged, thus lowering the costs of development. Further, this approach meant that the relation between the more entertainment-oriented fantasy scenario and the academic minigames was incidental, so there didn't need to be intensive design or curricular work to integrate the meta-activity with the academic tasks. The educational and entertainment genres were thus hybridized, but in a way that kept them essentially separate domains of activity.

As the educational software industry matured through the 1990s, the groundbreaking approaches of educators such as McCormick and Davidson were converted into an industry model that is more formulaic than revolutionary. Both McCormick and Davidson left the companies they helped create, and both cited differences with executives who came to run these companies and who focused on short-term corporate earnings. The design of games was systematized into a formula and established a market niche called *edutainment,* a label that McCormick "abhors." Davidson explained that she and her husband "had differences of opinion with the new owners over matters of company goals and values." Prompted further, she explained: "Companies need to be purpose-oriented as well as profit-oriented. Many media companies that create movies, television programming and software are saying that you just can't run a business without compromising on standards, but I don't agree" ("A Conversation" 1997, 25). In my interview with McCormick, she was even more direct, having been forced out of TLC early in its history.

I sold every share of stock. I wanted nothing to do with it. I sold all my stock for a dollar a share. When it went to sixty-five, I lost thirty million dollars making that decision. And I don't regret it. . . . They made it impossible to transform education alongside making huge profits by doing the same little programs over and over. Eventually that led to the industry crumbling because it didn't deliver on the promise of creating a resource that assures all children can learn what they need. TLC didn't make any transformative products after that, even though technology capabilities leaped forward.

Market demands converted McCormick's constructivist educational philosophy and egalitarian goals of reaching the technologically disenfranchised into a way of profitably delivering curricular content to middle-class families. This transformation is an indicator of how media content is inseparable from the economic conditions in which it is produced and circulates. Even when products are designed with the hope of transforming socioeconomic relations, market responsiveness means that these products often succumb to the inertia of established cultural categories, market segments, and social distinctions.

In the 1980s, educators with high ideals founded new companies, and these companies distributed new technology and products to a small market of like-minded educators and computer aficionados. The 1990s saw the proliferation of PCs, the consolidation of software industries, and the emergence of a consumer market in family-oriented software. Instead of being sold at specialty computer and hobby shops, most children's software was being sold at superstores such as Cosco, Walmart, CompUSA, Toys 'R' Us, and Office Depot. Career CEOs pushed aside company founders, and by the end of the 1990s, the children's software industry had largely consolidated under two conglomerates, one headed by Mattel and the other by media industry giant Cendant. Only a year after Mattel bought TLC for $3.5 billion, it sold the company to a turnaround specialist, Gores Technology Group, for $27.3 million. The latter in turn sold TLC to Riverdeep for $60 million in 2002 (Pham 2002). In 1999, Leapfrog Enterprises introduced a new hit product, the LeapPad, a platform that integrated physical books with digital interactivity (Helft 2008). Although the LeapPad and subsequent LeapFrog products in some ways injected new life into the software market for young children, the innovations were largely in hardware rather than in educational content. The market for children's software was increasingly pushed into a narrow niche of basic literacy and early childhood education, with elementary-age kids quickly moving on to the more compelling offerings presented by video games and free content on the Internet. The market for PC-based educational software plummeted from a high of $498 million in 2000 to $152 million in 2004 (Richtel 2005).

The 1980s and 1990s saw the networks of technology, people, and capital for children's software extend far beyond the boundaries of the original small-scale market made up of progressive intellectuals and technologists. A larger market, mainstream retail, and more resource-intensive forms of technology ironically led to the demise of what many have felt to be quality products. The greater production expenses associated with cutting-edge technology and graphics also means that there are higher economic barriers to innovation, lending more inertia to the conservative tendency. In contrast to the early experimental years, companies are now focused on upgrading graphics and sound and on developing content in established formulas rather than on developing new models for interaction or game design. The development process has shifted from a reformist educational impulse to a more market-driven emphasis. One developer described to me the climate in the early 2000s. Unlike for the early TLC products, she said, "the impetus for these games comes from marketing. This is retail marketing, not school marketing." She described how the results of market surveys and shopping-mall intercepts define the initial parameters for a new product, and then designers are given a budget, time frame, and schedule. "The budgets have been shrinking. The calendars have been growing shorter, so there's a lot of pressure to turn things out quickly." Explaining how these market pressures limit content development, she commented, "It's been quite honestly very frustrating to people in this company to have smaller budgets, less time. There has been a very great emphasis on reusing assets. Some of this makes good sense and some of it is jut cost cutting and corner cutting. " But, she continued, "I'm amazed at what our learning specialists and producers can do with shorter time and shorter budgets. They still turn out good products. But the market pressures are there—for example, for these grade-based products. Well, if you're going to do something for fifth grade, and you're going to cover the major content areas, you're not going to do any of them in any kind of depth."

The content of educational software has over the years reflected these shifts in marketing and distribution. Although the look and feel of learning software have benefited from higher-end graphics and more sophisticated multimedia PCs, the content has grown increasingly systematic, and marketing is based on brand recognition and achievement anxiety rather than on innovation in design and depth in content. This retrenchment ironically began to happen at a time when the technology was starting to achieve the potential to support much richer forms of content than those available in the early years, and video games were capitalizing on these new capabilities. Products with easily represented marketing "hooks"— such as a licensed character, an established brand, or the claim to transmit strategic cultural capital—became easier to disseminate in a consumer market than products with more open-ended, complex, and multireferential goals. These shifts in orientation also related to the quickening climate of the "new economy" of the late 1990s, where notions of competitive success were intimately tied to efficient new technologies that amp up performance. The products left standing in the academic genre play on parental anxiety about whether their children are "competitive" not only in terms of acquiring the cultural capital of school subjects, but also in terms of having an achiever's identity and a competitive stance. This orientation feeds into a structure of participation with new media that focuses on academic success and competitiveness.

An early product, *Oregon Trail,* placed academic knowledge in a meaningful context of historical simulation in a way that departed from the standardized curriculum and testing demands of most schools. As children consider how best to manage rations and supplies and to proceed along a simulated journey with real-world referents, academically relevant content is mobilized as one relevant component of decision making. There is no hierarchical assessment of achievement based on realization of a single correct outcome. The products that became dominant in the 1990s, by contrast, put social distinction and assessment back in, which was key to their marketing success. In efforts to create the self-esteem and identity of an academic achiever in players, such products repetitively applaud children for getting the "right answer" to academic problems that are unrelated to the fantasy adventure scenario that is presumably the "fun" part of the activity. These successes are framed and tallied like grades in school. In marketing materials, parents are told that these products will ensure that their children will internalize the dispositions and cultural capital necessary for competitive academic success. Contrary to McCormick's efforts to focus on learning rather than on education, one sector of the industry has found that achievement, in the form of school success, is the most easily marketable package for academic learning software.

Consuming Achievement

Grade-based educational software appeals to parents' desires for wholesome, creative, and interactive play for their children that will give them a leg up on subjects covered in school. These products appeal to the middle-class parenting approach that Lareau (2003) describes as "concerted cultivation," in which children's time outside of school is occupied with "productive" forms of play. Unlike entertainment-oriented games, which are marketed directly at children on television and in gaming magazines, educational software is marketed toward parents. Ads for children's software ran in *Family PC* until it became defunct in 2002, providing us with a window into how these products were marketed. One ad for Knowledge Adventure's *JumpStart* software series (figure 2.2), running in the December 2000 edition of *Family PC*, sets up an unambiguous relation between the products and academic achievement. Still wearing her backpack, a blond, school-age girl dressed neatly in white knee-high socks, Mary Janes, and a red skirt stands with her back to you (your child here) and clutches a school worksheet. The sanitized space of the large kitchen and the girl's appearance code the home as white, suburban, conservative, and middle class. The girl faces a refrigerator covered with assignments red inked with gushing teacher notes: stars, "Good Work!" and "Excellent!" The backpack, the school assignments, and the voice of assessment are represented in a central role in the intimate sphere of the home. A drawing of Mom, posted in the visually prominent area at the top left, hails the parent in charge of children-related purchases. She is a smiling mother with curly blond hair and rosy cheeks.

The ad copy describes the current concern with self-esteem and identity in promoting academic achievement. "When kids succeed, they feel confident. When they feel confident, they succeed. This is how *JumpStart* works. And why so many parents think it's the best learning software you can buy." In contrast to the other ads in this campaign, this one features a girl and has ad copy that specifically poses self-esteem issues. Together with the girl's assertive posture—hands on her hips, head tilted upwards—the ad's text implies that the software will address confidence issues that plague girls in academic achievement. The software claims to provide a jump start for children stalled in the academic rat race, mobilizing the metaphor of "education as a race" that dominates the culture of competition in U.S. schools (Varenne, Goldman, and McDermott 1998, 106–115). The ad campaign's tagline, "She's a JumpStart Kid, all right," is subtly crafted to imply a status distinction from other kids, the perpetually stalled failures who don't use this software. The phrase *all right* is a reassuring confirmation of parents' conviction that their child is inherently smart

Figure 2.2
Advertisement for *JumpStart 1st Grade*. Reproduced with permission from Knowledge
Adventure, Inc. JumpStart is a trademark or registered trademark of Knowledge
Adventure, Inc.

and deserving of this status. Your child, too, may be deserving of greater recognition of success than she is currently receiving.

The ads for toddler and preschool software titles target increasingly younger children and promise to fill a child's "leisure" time with the competitive logic of academics. Another ad in this campaign features a smiling, sleeping boy in a bed covered in books with titles such as *Ships, Vikings,* and *The Stars.* The books are even tucked into his bedsheets, replacing the stuffed animal so iconic of childhood attachments and imaginings. The boy is presumably integrating academic content into his dreams. In yet another ad, a *JumpStart* kid, again with a backpack on and lunchbox in hand, is waking up a bleary-eyed father at the crack of dawn. These kids have internalized the disciplines of schooling and are thus reluctant to take their backpacks off and eager to get them back on. They are represented as identifying deeply not only with academic content, but with the aggressive, forward, and upwardly competitive stance of academic success.

Corporations market software to parents as a vehicle for academic success, and parents in turn market academics to their kids as an entertainment activity. Parents can mitigate their sense of being pushy in their achievement orientation with the reassurance that their kids are having fun. Academic content gets integrated into children's self-identities and pleasures. The ads for the *Math Blaster* series feature children in moments of ecstatic play, swimming or playing superheroes, with thought balloons describing the mathematical significance of their play (figure 2.3). A tiny caped crusader speculates, "If I fly 90 miles an hour and the earth is 24,902 miles around, can I still get back home for breakfast?" "Must be the Math Blaster®," suggests the ad copy below the picture. "Software that gets your kids into math. And math into them." As Loyd Rieber, Nancy Luke, and Jan Smith argue in their review of learning-game design philosophies, this approach looks at "fun" as an extrinsic motivator of learning, rather than working to support learning that is intrinsically motivated. The focus on entertainment as a motivator tends to "designate the role of games as a form of educational 'sugar-coating'—making the hard work of mathematics or language arts easier to 'swallow' " (1998, 5). This approach toward extrinsic motivation relies on a behaviorist model of learning (Engenfeldt-Nielsen 2006).

It is not sufficient for children to perform well academically; learning needs to be fun, and children need to *love* it. "I Love Reading!" "I Love

Figure 2.3

Advertisement for *Math Blaster.* Reproduced with permission from Knowledge Adventure, Inc. Math Blaster is a trademark or registered trademark of Knowledge Adventure, Inc.

Spelling!" trumpet the titles of a learning series from Interactive Learning, the ads for them adorned with the faces of wide-eyed, smiling children. Learning for pleasure must infuse kids out-of-school lives. Like children's literature and educational toys, educational software holds forth the promise of learning that is tied to school success, yet is freed from the dusty, boring school atmosphere—learning that promises joy, delight, engagement, and identification. A token African American child graces the cover of one of these titles, surrounded by a sea of white faces. The white, middle-class marking of these ads and the hefty prices of the products and computers indicate that the market is for families that are seeking to maintain middle-class status or are aspiring to upper-class status, not for the racially diverse and disenfranchised populations that were the target of McCormick's efforts. The progressive philosophy of "learning through play" has been transformed into a more conservative agenda of "achievement through play," an important shift in these products' orientation.

If parents pick up a software box at Cosco, Toys 'R' Us or Comp USA, they can learn a bit more about the software contents. Most software boxes feature a front flap that can be opened, revealing content domains and providing screen shots of different games screens. For example, the *Jump Start 2nd Grade* box cover features the product's key elements: the title, the green frog character that guides the adventure, the company name, the target age, the tagline "There's No Stopping a Kid with a JumpStart!" and, crucially, a seal attesting that *JumpStart* is "#1," with more than three million copies sold (figure 2.4). Successful products such as *JumpStart* push their brand as a central marketing vehicle. Parents are enlisted as believers in the "JumpStart family" of products that range from titles for toddlers to titles for children in the upper elementary years. At the top of the box, the company, Knowledge Adventure, is associated with the tagline "Discover. Learn. Excel," tracing a three-point progression from the child-centered ideal of discovery and exploration to learning and identification and finally to competitive success. As a visual genre, these boxes draw from the representational styles of children's picture books and the parent-friendly animation of the nonviolent PBS variety: bright colors and cute, wide-eyed, anthropomorphized animal characters with big smiles and big heads. Although not visually central, certain elements in the cover serve as code for "school": the ruler notches in the title bar, numbers, a plus

Figure 2.4

Box cover for *JumpStart 2nd Grade*. Reproduced with permission from Knowledge Adventure, Inc. Jump Start is a trademark or registered trademark of Knowledge Adventure, Inc.

sign, and the words *noun* and *verb* as part of the background scene. The box translates curricular content into the aesthetically pleasing vernacular of children's edutainment, much as we find alphabets and numbers adorning infants' bedding and toys.

Opening the front flap, a parent sees the claim emblazoned across the top: "A Full Year of 2nd Grade in an Exciting Adventure!" Below that are screen shots of each activity, describing the academic content involved (figure 2.5). For example, "Ice Cavern Math" teaches multiplication tables, and "Save Our Universe" teaches about the solar system. The back of the box gives a list of what "kids learn" and a checklist of "what you get": "Over 80 Skills Taught," including "Simple Multiplication" and "Social Sciences"—in other words, "a complete 2nd Grade curriculum." This "grade-based system that grows with your child" packages learning as the ability to progress along an atomized set of tasks as defined by the basic components of a school curriculum. A small girl is pictured sitting on the progressive steps of this "learning system." There is also a photo of this same smiling blond child, accompanied by parental testimony regarding how "Amanda" is making so much progress in her schooling. The software's technical features outlined on the back of the box center on parental control and guidance in the context of a wide-ranging set of skills and activities (figure 2.6). The game has "adjustable difficulty levels" and offers a "parent's progress report." In contrast to other forms of technical engagement that stress the empowerment of the child, this product highlights the function of technology to structure and monitor behavior.

The defining characteristic of the academic genre is that titles are age graded, relying on an "ages and stages" approach to development, oriented toward maturation along adult-defined measures. Cute, innocent characters, primary colors, and content tied to school grades marks this product as one that kids older than ten will not identify with. Marketing is directed at parents who want their children to succeed at school. Efforts are made to include girls in the framing and marketing, but the racial and class markings emerge as white and middle class, muted somewhat by the use of animal characters. Distribution is through mainstream mass retailing that touts the product's "number one" popularity. These characteristics define the academic genre of software at the level of design, marketing, and distribution. A specific case study of the content of one game typifying this genre shows

Figure 2.5

Inside the box of *JumpStart 2nd Grade*. Reproduced with permission from Knowledge Adventure, Inc. Jump Start is a trademark or registered trademark of Knowledge Adventure, Inc.

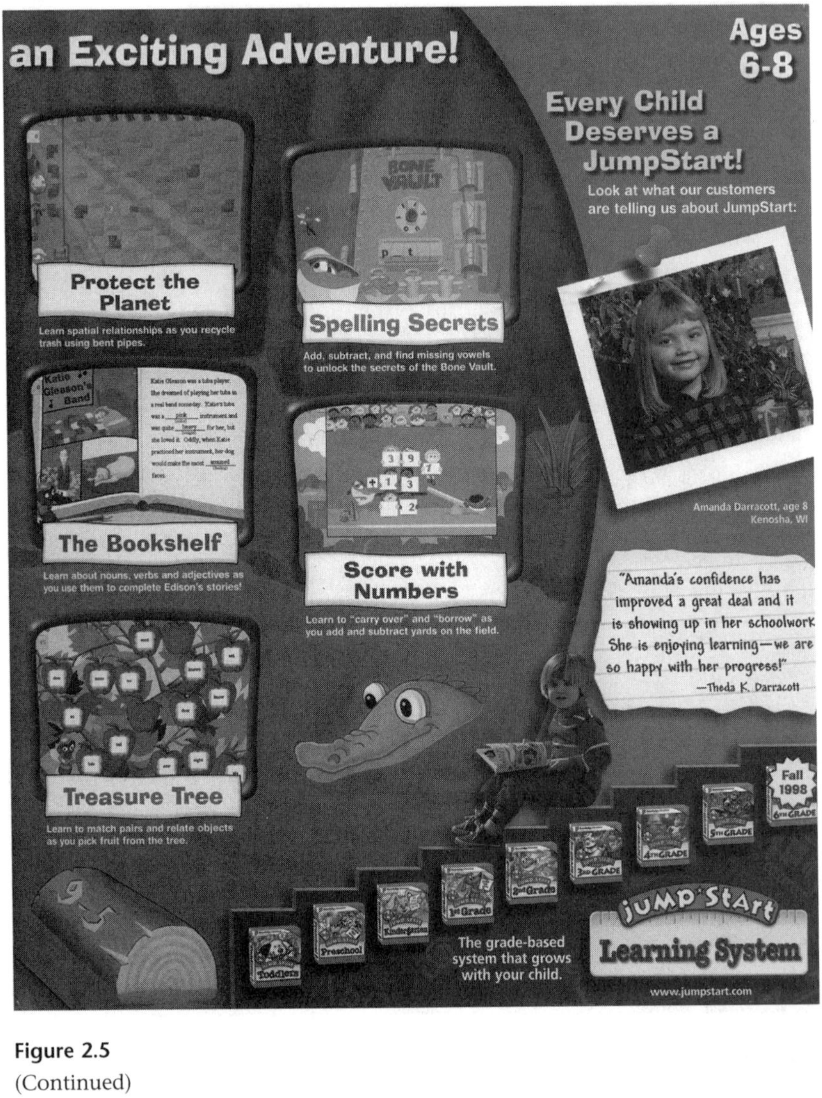

Figure 2.5
(Continued)

Figure 2.6
Back cover of box for *JumpStart 2nd Grade*. Reproduced with permission from Knowledge Adventure, Inc. Jump Start is a trademark or registered trademark of Knowledge Adventure, Inc.

the way it was engaged with at 5thD clubs during the time I did my field-work and provides further insight into the genre as a whole.

The Academic Genre and *The Island of Dr. Brain*

Software from companies such as Knowledge Adventure and TLC have been a mainstay of the 5thD clubs. In the late 1990s, when I was conduct-ing my fieldwork, the clubs still ran copies of games that McCormick had been involved in producing, such as *Gertrude's Puzzles,* on old Apple IIs, and as they upgraded their machines to more sophisticated models, they purchased and utilized more recent titles. The grade-based systems such as *JumpStart* were yet to make an appearance, but some available products relied on a similar adventure-puzzle format. During my fieldwork, a new game, *The Island of Dr. Brain,* was introduced, which was played from a CD-ROM and had more sophisticated graphics than the earlier adventure-puzzle games being used at the club, such as *Gertrude's Puzzles. The Island of Dr. Brain* is one in a series of three *Dr. Brain* titles published by Sierra Online, a company known for its adventure-game titles for adults and kids alike. Here I use *Dr. Brain* as a case study typifying the academic genre and how it resonated with the 5thD's orientations toward academic play, but also created problems by inviting a focus on progress through the game rather than deep mastery of content.

According to one of the undergraduates who volunteered at one of the 5thD clubs I observed, *The Island of Dr. Brain* is a "Mensa-like" game involv-ing a series of puzzles that are reminiscent of IQ or school tests. These puzzles are embedded in a fantasy, role-playing scenario, where the player is working as Dr. Brain's lab assistant to recover a special battery from his secret island. The game's preplay sequence includes an animated movie in which the player is introduced to the benevolent, goofy, but demanding patriarchal figure Dr. Brain, a predictably white-haired, lab-coated fellow, and given a set of instructions for how to play the game. The player as lab assistant is then deposited at the entrance to the island and encounters the first of a series of puzzles and intelligence tests. All of the tasks in the game are organized around science, math, logic, and other "brainy" subject matters. For example, in order to get into the island, the player has to solve a polynomial puzzle that involves placing puzzle pieces correctly on the door to the island. In the entrance chamber, the player then clicks on a microscope to solve an algebra problem to sort some microorganisms along x and y axes, a number series problem to open a sarcophagus, and the Tower of Hanoi puzzle to open the door to the chamber. The game proceeds in this manner through different scenes with embedded puzzles: a tropical forest, a bridge, a volcano, a village, a hut, and Dr. Brain's laboratory. At the end of the game, the player is treated to a lengthy animation sequence, given a final "report card" of how she or he did on the different problems, and encouraged to play again at a higher level of difficulty.

The look and feel of the game are entertainment oriented and graphically sophisticated, referencing the wacky "mad scientist" tropes in educational television programming (MacBeth and Lynch 1997). Peripheral elements such as humorous dialog boxes and animations mute the game's didactic tone. Although the look and feel have been designed with entertainment idioms, the game's content is essentially academic; the "brainy" content is thus packaged as a cool and fun domain. The game follows a coherent adventure story line, but the content of the puzzles are largely incidental to the fantasy scenario: a spectrum analyzer puzzle may be the condition for opening a doorway or a synonym puzzle a way of getting a basket of apples. The fantasy scenario furnishes a narrative coherence and goal orientation to a series of otherwise unrelated problems. This dual structure also corresponds to different forms of knowledge. The exploratory elements and fantasy scenario are tied to game-specific forms of knowledge, including the particular narrative meanings of *The Island of Dr. Brain* as well as the features that relate to other adventure games and entertainment media (i.e., what to click on in a given scene, the sequence of puzzles, the "mad professor" fantasy narrative, etc.). The knowledge embedded in the puzzles (i.e., math, science, and other academic content) is meant to be "transferrable" to school contexts. This lack of relation between the fantasy scenario and the puzzles parallels the distinction between the game's entertainment look and feel and its academic content.

The Island of Dr. Brain is functionally relatively simple, but it relies on multimedia to capture the player's attention. Both the preplay and ending sequences involve extended animation scenes, and the game throughout makes use of color graphics, animation, and sound that were considered cutting-edge in the mid-1990s, when I was observing the use of the game. Each ambient scene in which the puzzles are embedded are functionally much like an animated storybook, with "hot" areas that trigger either an animation, a dialog box, or a puzzle. Although players can explore the storybook scenes in an unmonitored and open-ended way, these exploratory moments are coded as silly and functionally inconsequential, transitional "downtime" between the "real work" of solving serious academic problems. The game's narrative trajectory is sequential and single track; although the player can revisit previously solved puzzles, there is only one pathway through the island, and getting to any given puzzle is contingent on solving the puzzles preceding it. The puzzles are similar to a standard test or workbook, based on consideration of a limited set of answers to a problem with clear right and wrong answers. The game allows for different levels of difficulty and keeps track of how many puzzles are solved correctly, which corresponds to the report card presented at the end of the game.

The game's other functions include a scorecard that can be accessed at any time, the ability to save a game and set the level of difficulty, and the ability to revisit previously solved problems. Players can also click on a "hint watch" that triggers a dialog box that gives them clues on how to solve a problem. The number of "charges" in the hint watch is limited, restricting the amount of help kids have access to. Beyond these features, there are no other ways for a user to engage with the game's technical functionality. Players' ability to change technical parameters is restricted to navigation, assessment, and monitoring functions. Although framed in the context of play and exploration, the game's overall structure parallels features of school learning, where students submit to the authority of the teacher (Dr. Brain) and of canonical knowledge embedded in a series of discrete tasks, which are thought to build on each other in a necessary sequential order. Accomplishment of each task is awarded by "grades" (gold and bronze plaques) and tallied into a final score at the end of the game. The game even provides extra-credit points for kids who solve additional problems. Unlike school, however, negative assessments are muted in encouraging messages such as "Try again!" Game outcomes and grades do not perform the same sorting functions as school assessment, so the game is an arena for boosting academic self-esteem.

Dr. Brain in the 5thD

Despite its school-like properties, *The Island of Dr. Brain* was remarkably popular in the 5thD in the 1990s among Young Wizard's Assistants and kids willing to use free passes to play it. Overall, kids responded positively to the eye-catching graphics and animations, and oriented quickly to the game's linear goal structure. For all of the kids we observed on tape, puzzles at the novice level were doable with some help; the game skirted the edge of kid expertise, inviting productive collaboration with adult participants. During the period when we were observing, the game had just been introduced to the site on a trial basis and was not included in the site's activity system. Because of this special status and because of its cutting-edge graphics, it attracted the attention of researchers, undergraduates, and kids. The undergraduates' field notes describe the game as "challenging," "impressive," "thought provoking," school-like, with "obvious" educational value. One undergraduate also points to the "diversity of tasks" as important in making the game interesting and appealing to different kids. During play, the kids' descriptive discourse about the puzzles focused on whether the game was "hard" or "easy." This kind of discourse contrasts to discourse about other forms of gaming, which runs along a spectrum from "cool" to "boring." When playing *Dr. Brain*, kids used the term *cool* only to describe elements of the exploratory scenes and animations, never the puzzles.

Researchers at 5thD initially thought that the game might provide a useful vehicle for comparative study of engagement with similar cognitive tasks across school and the 5thD; the game included tasks often encountered in school, but here framed by an entertainment-oriented structure and the 5thD's unique social context. We videotaped kids playing the game during a period of about a month in the fall that the game was introduced. The corpus of videotapes and notes about the playing of this game are larger and more consistent than for most other games because of this focused attention and because the game requires multiple days of engagement to complete. Many of the simpler games have tasks that can be completed within one club period or do not have a linear goal orientation and so do not invite sustained involvement with the game over time. With *Dr. Brain*, we have a rare record of a number of kids engaging with the game across multiple sequential days. After some time evaluating the game, however, the researchers at the site decided that it was actually a poor vehicle for evaluating the particularities of the learning processes in the 5thD, though the game continued to be included at the site. They felt that the game tasks are atomized in a way that worked against the deeper intellectual engagements that the 5thD encouraged, and that the game fostered a competitive orientation toward achievement that was not part of the 5thD ethos.

The game called forth the latent tension between the 5thD's educational philosophy and mainstream schooling's educational philosophy. This tension manifested itself most obviously in the difficulty of recognizing and managing social markers such as "smartness" and "success." In the field notes and videotapes of play, three salient dynamics emerge that define the academic participation genre. These dynamics relate to the tensions regarding the value of the game in a setting such as the 5thD. One dynamic is the way in which entertainment and education idioms are incorporated into the game as fundamentally disjunctive forms of engagement. Another is the tension between the orientation of kids, who want to get through the tasks in as expedient a manner as possible, and the orientation of adult helpers, who try to get kids to understand the nature of each problem raised in the game. The third is the way in which the game invites a competitive orientation through explicit achievement recognitions in the game itself as well as through knowledge and achievement displays by the kids. After describing these three dynamics, I present a case study of one boy who performs the achievement genre of participation in relation to this game and thus exacerbated these latent tensions in the 5thD system.

Marginalizing Entertainment

Problem-solving strategies for *The Island of Dr. Brain* involve two different modes, which correspond to the game's oscillating structure. When engaged in the click-and-explore scenes, kids will click around and solicit feedback from the scene, eventually hitting on the right element that will make a puzzle pop up. Although the animations and humorous messages that pop up when a player clicks on the "wrong" elements are amusing for a while, even relatively short periods in this mode invite a sense of frustration at not making progress along the game's sequential logic. Problem solving in these scenes is based purely on guessing and trial-and-error strategies, unless a local expert is present who already knows where to click on the scene. For example, in one tape, two kids are trying to figure out what to do in a room in Dr. Brain's lab. It takes them about twenty seconds to pick up a cartridge, but then they are stuck for about a minute trying to find where to put the cartridge. After clicking around in vain during this time, the kid controlling the mouse screams, "AAAAH!" in frustration, clicking randomly all over the screen, before being calmed down by his companion and finally figuring out that one needs to click on a specific part of the robot to get to the programming puzzle: "Yes, I got it!" More commonly, engagement with the transitional scenes is an unproblematic and brief break from the puzzles. Kids will click randomly around the screen and will fairly quickly hit the object that will lead them to the next puzzle. In breakdown situations, however, when kids get "stuck" in a particular scene, they feel a great deal of frustration and a sense of injustice at not being able to move ahead. One case involving an extended breakdown sequence called forth this kind of frustration.

The tape shows Cathy, a twelve-year-old, playing the game for the first time, moving quickly through the puzzles, and solving them with ease. In other words, she is having no difficulty in navigating the game's narrative or problem space. At a certain point, however, she runs into some difficulty. She finishes a problem that involves manipulating some microorganisms under a microscope. When she exits the problem, she is presented with a slip of paper. She inserts the piece of paper into a sarcophagus, which is also on the screen, and is presented with a number series problem, depicted as the lock to the sarcophagus. When she solves the problem and exits the puzzle, however, nothing happens, and she declares with dismay, "It didn't open." She tries clicking on different parts of the sarcophagus, but nothing happens. The undergraduate working with her suggests, "Maybe you messed up the first one," and Cathy goes on to repeat the microscope problem. She again solves the problem with minimal effort, exits, but nothing happens. She does the microscope problem again. Nothing happens. She clicks on the sarcophagus and gets a humorous message that doesn't help her get the lock open. She does the microscope problem again. Nothing happens, and she tries again, declaring, "Oh! This is making me mad." She does the problem again, trying to click on different parts of the puzzle, but to no avail. She does the problem again and again. The following exchange then ensues, involving both the undergraduate and the site coordinator:

C = Cathy
UG = Undergraduate
SC = Site Coordinator

1   UG: Are you stuck?
2   C: I'm mad 'cause I put the card in there, but then it won't do it. Watch.
          (Solves microscope puzzle.)
3   UG: So if you mess up, it doesn't accept it?
4   C: But I didn't mess up. (Solves the microscope puzzle again.)
5   UG: So you have to finish this in order . . . hmm, for it to work? Wait, it's . . .
6   C: Watch, look at this. (Clicks around main screen and gets humorous mes-
          sages and animations.) I got the card, and then I put it over. (Clicks around
          screen and gets various animations. Solves microscope puzzle.)
7   C: (Calls site coordinator over.) I played this game at least six times, but it
          won't let me. Watch. Watch. I'll play it. (Solves microscope puzzle.) Watch. I
          did that. Watch this. I go to here, and I don't get anything.
8   SC: Yeah. Yeah. 'Cause you've already done this part. This is the same game,
          right?
9   C: Yeah.
10   SC: Yeah, it doesn't have anything to give you, you've gone backwards.
11   C: Oh.
12   UG: So you just keep on going.
13   SC: Yeah.
14   C: So what do I do? (Clicks on door to exit the screen, but game doesn't let
          her exit.)
15   K: The door won't open, though.
16   SC: How did you get back here? How did you get back here?
17   C: It started me all over again.
18   SC: I've never seen this happen. I don't know. See if maybe you can open
          the sarcophagus. I don't know.
19   C: (Clicks on sarcophagus, and it doesn't open. Looks up at SC.) How do I
          get out of it?
20   SC: (Pause) I don't know. You can just quit, but don't save it.

This series of events is frustrating, with Cathy repeatedly solving the
problem as dictated, clicking on all of the resources available on the screen
(lines 7, 14), and soliciting help from two adults (lines 2, 7). She gets
increasingly angry and insists that she hasn't "messed up," solving the
problem yet another time to demonstrate that she is doing things correctly
(line 4). Even the site coordinator is baffled as to why the game is not
proceeding along its narrative trajectory (line 18). Cathy eventually has to
start the game over from the beginning. The hints and other feedback that
the game provides are designed only to buttress the narrative fantasy,

causing objects on the screen to animate, giving hints on how to solve the problem that she has already completed or posting funny messages that give no hint as to the game's underlying functionality that would allow the player to proceed to the next puzzle.

If Cathy had clicked just a few centimeters off in one direction, directly on the lock of the sarcophagus as opposed to on the casing, she would have moved on to the next sequence in the puzzle. The game, however, provided no clues that would identify a particular part of the sarcophagus as an activation point. Finding where to click was not designed as a significant bit of knowledge, and there were no hint or help functions to instruct Cathy about a way out of her predicament. The exploratory scene was designed based on the genre of entertainment, focused on open-ended exploration, which was not the genre of participation that Cathy was orienting to, thus the game experience generated frustration for her.

Unlike in this sequence, both the kids and the undergraduates we observed considered engagement with puzzles, even over extended periods of time, "productive." Kids would pursue a single problem for more than ten minutes without exhibiting comparable frustration. This difference is consistent across all the instances of play we have in our observational record on the game. The puzzles involve small, self-contained tasks with fairly explicit instructions and recourse to the "hint watch" that displays a partial solution. When the puzzle first comes up, a series of pop-up windows states the instructions for the game, and the game provides ongoing feedback on the user's actions while he or she is working on a puzzle. For example, a hidden-figures problem gives feedback such as "you're getting warm" as the player clicks around the screen. There is no such feedback in the transitional scenes like the one with which Cathy struggled. An accompanying book called the *EncycloAlmanacTionaryOgraphy* provides content-based information that helps solve puzzles, although it was not in use at the 5thD at the time I did my fieldwork. Solving puzzles is the rubric under which the game evaluates players and provides recognition of success in the form of bronze and gold plaques.

Unlike the puzzles and academic content, the fantasy scenes are presented in decontextualized and lightweight ways, involving a pastiche of styles and references that draw from "mad scientist" tropes, tropical island fantasies, and jungle adventure stories. No additional information is provided on how to navigate these scenes, and they are clearly marginal to the game's overall goals. The game does not keep records of where players have clicked on the transitional scenes, and when the player clicks on the right place to move forward, there is no explicit recognition of "success" other than the transition to a puzzle. Undergraduates acknowledge in their field notes that they resisted "just giving kids the answer" to problems, but they had no problem with the sharing of knowledge that is specific to the fantasy scenario, such as where to click on the scene in order to get to the next puzzle. In other words, both undergraduates and kids recognized that the dominant logic of the game is in the academic genre and that the entertainment elements are peripheral and meant to be bypassed quickly. The transitional scenes are visually attractive and occasionally amusing, but they are functionally inconsequential for the recognition of achievement and score keeping. The remainder of the chapter focuses on the problems as the central feature of playing the game.

Playing the System

Although *The Island of Dr. Brain* embodies the academic genre for both design and engagement, the game doesn't necessarily invite deep engagement with the problem and academic content. Although all of the game's puzzles exhibit some kind of "brainy" content—chemistry, math, art appreciation, and so on—many can be solved tactically based on the logical consistency of the problem domain rather than with recourse to the broader domain of knowledge. Players rarely discuss the academic content (e.g., the nature of algebra problems, what a microchip does, what a dominant and recessive gene is). In many cases, the puzzle is self-explanatory or recognizable from other game or test situations: a hidden-figures puzzle, a matching game, a word search puzzle, a number sequence puzzle, a jigsaw puzzle, or magic squares. In other cases, the problem's structure emerges after clicking on some elements of the puzzle and through trial and error. For example, one problem asks for solutions to an algebra equation to change the lines on a set of x and y axes. After clicking the numbers in the equation, kids at the club quickly figured out the relationship between the equation and the line on the graph and soon solved the problem. In still other cases, the problem's structure is more opaque and invites recourse to the hint watch, a rereading of the instructions, or enlistment of help. Regardless of the difficulty of recognizing the general nature of the task, overall engagement time with the puzzle is almost always dominated by trying to figure out the problem's structure. For example, although the idea of hidden figures is almost immediately recognizable, the details of how a player selects an answer and what the hints mean (such as "you're getting warm") require explication.

In her book *Plans and Situated Actions: The Problem of Human/Machine Communication*, Lucy Suchman (1987) analyzes the interactions between people and a function-loaded copier with a computerized, interactive interface. Drawing from ethnomethodological frameworks and previous studies of face-to-face interaction between people, Suchman posits a basic asymmetry in the interaction between people and machines: "[P]eople make use of a rich array of linguistic, nonverbal, and inferential resources in finding the intelligibility of actions and events, in making their own actions sensible, and in managing the troubles in understanding that inevitably arise" (180–181). By contrast, machines "rely on a fixed array of sensory inputs, mapped to a predetermined set of internal states and responses" (81). Through close analysis of interactional sequences of copier use captured on videotape, Suchman demonstrates some of the interactional outcomes of this asymmetric communication. Of particular interest are breakdown situations, where the machine's responses, as anticipated and preprogrammed by systems designers, are not appropriate to the users' specific needs and situations, so human and machine are unable to engage in mutually intelligible action. Suchman documents many instances when the user and machine have been unable to orient to a task, such as making double-sided copies of a document, because of the mutual unintelligibility of their respective actions. Users often develop workarounds that enable them to overcome the limits of the copiers' procedures.

Unlike competition with other people, competition with a computer is about anticipation of fixed interactional responses. The conditions for achieving the desired end state may be dormant, but are already programmed into the system. In this form of competition, mastery is not about responding flexibly to an unpredictable opponent who may be indexing a wide range of possible strategies and skills. Rather, competition with the game is about orienting to a set of consistent and algorithmic responses (e.g., building an office unit in *SimTower* increases population by *x* amount) or to a predetermined narrative sequencing (e.g., a player clicks on an apple tree, then progresses to the next scene). People are able to find innumerable ways to outwit the machine by means of their incomparably richer indexical resources (Ito 2006a). For example, a child can find game-playing tips in a magazine or on the Web, work in collaboration with knowledgeable peers, or simply restart the game when the going gets too rough. With games based on singular goal orientations, the interactional asymmetry between human and machine invites creative workaround and alignment with the rigid conditions set up by the game for the player to win. The game stands in for the designers' educational goals, but is not able to reproduce them in a situationally responsive way. In practical terms, this often means that kids orient themselves to the precise inputs necessary for moving ahead in the game rather than to the actual content (especially if educational) to which these inputs are supposedly tied.

In the 5thD, kids are also able to call on the adults on site for help. As in most explicitly educational settings, the adults feel obliged to withhold solutions to problems in the service of the kids' intellectual development, but the kids want to be given the answers in order to beat the game. In other words, for kids, winning matters more than how they play the game, especially when playing against a stubborn machine. Computer games don't account for or acknowledge how kids play the game beyond the specific inputs at the interface, and thus they can invite a focused orientation to achieving these formal conditions for winning at the expense of understanding or mastery of the process of play. This orientation is exacerbated by the fact that the problems in games such as *The Island of Dr. Brain* are fragmented and do not build on each other in any way, so mastering one problem does not help to speed a player through other parts of the game. Some elements of play in *Dr. Brain* are exemplary of this orientation. In one sequence of activity we recorded, Roger, a twelve-year-old, moves quickly through the various puzzles with the help of an adult. Then they arrive at a hidden-word puzzle, where they need to find and highlight words in a foreign language. Roger chooses French, and they proceed in finding the words.

R = Roger
A = Adult

1   A: It's in French. Hmm?
2   R: We have to know . . . oh, it's "au revoir." . . . Where's "au"? Help me find it. Look for an *a*.
3   A: Oh, I see it (pause).
4   R: Where? Just tell me.
5   A: You're hot, hot, hot, hot (referring to where Roger is moving on the puzzle).

6   R: Hot?
7   A: No (as Roger moves cursor). Hot. Burning. You're burning hot.
8   R: Oh, here it is.

In this sequence, Roger immediately demands help in finding the words—"Help me find it" (line 2). When the adult finds the word, she pauses, not pointing it out to him, to which he responds impatiently, "Where? Just tell me" (lines 3, 4). She compromises by giving hints, "You're hot, hot, hot," rather than tell him the answer directly (lines 5, 7).

The undergraduates in the 5thD work hard to apply the Vygotskian educational philosophies that they are learning in class to their interactions with the kids. They have been instructed to "provide the children with as little help as possible, but as much help as necessary to ensure that both the students and the children have a good time" (Cole 1997, 298). This rule of thumb for pedagogy in the 5thD is an application of Vygotsky's theory of the zone of proximal development, which posits that learning happens as an interaction between experts and novices engaged in joint activity. In their field notes, undergraduates describe their interaction and work to exhibit their application of these educational principles. One undergraduate describes how she provided information to help her partner move ahead in the task, but that the kid had a good basic understanding: "The first puzzle we had to solve was matching elements to their periodic table names. I went through the same steps he did, and when he figured out which element he wanted to match it with, I helped him by telling him what the Latin words were that the periodic table names were listed under. He figured out the abbreviations from there. He knew most of them on his own."

Another undergraduate similarly describes how she provided just enough help, but that the kid was doing most of the work in solving the problem: "Some of the language was very complex, and I helped him by telling him what the words meant. I would just give him the meaning of one of the words that needed to be replaced, and if he didn't know what the other words meant, I would give him pointers on their general meaning, or use them in a sentence, so he could figure them out for himself."

In contrast to this stance by adults, who feel it is important to provide the minimal amount of help, kids seem to have no reservations about giving the answers directly to other kids. In one instance recorded the day after the previous sequence with Roger occurred, a younger boy, Chris, is playing *Dr. Brain,* and Roger shows up to dictate exactly what to input, with no explanation of the actual process. Chris has just begun a new problem involving programming a robot to move through the laboratory and pick up a silver key. Roger immediately declares his expertise and then tells Chris what to do. The remainder of the problem-solving sequence, until they are able to get the key, is exclusively about Roger dictating operations and Chris inputting them.

R = Roger
C = Chris

1	R:	All right, go in.
2	C:	Here we go.
3	R:	I'm very good at this.
4	C:	There was the earthquake. Now what do we want to do?
5	R:	Take the purple one.
6		(Chris picks up purple chip.)
7	R:	Put it in the back of his head.
8	C:	Right there? (Puts it in robot's head.)
9	R:	Yeah. OK, let me do this.
10	C:	We know how to . . .
11	R:	I'm really good at this.
12	C:	OK, hint watch. (Hits "hint," which turns position indicator on.)
13	R:	Hint watch. OK, robot position indicator.
14	C:	OK we want to go—
15	R:	Forward.
16	C:	Forward.
17	R:	Forward, left, forward, forward, forward, forward, stay, stay, stay. Pick it up. Pick up. Now right, right, forward, forward, forward, right, forward, forward, drop. Now run, I mean go. Do go.
18		(Chris inputs program as Roger dictates. Then he runs program, and robot gets a silver key.)
19	R:	There he goes. OK.
20	C:	Silver key.

In this sequence, Roger begins by telling Chris how to activate the puzzle (lines 5, 7) and then dictates the answers as Chris inputs them (lines 15, 17). The two kids' jointly formulated goal is to get through the problem as quickly as possible rather than to explicate content knowledge for Chris, who blindly follows Roger's instructions. Both of these sequences, although based on different priorities between kids and adults, are outcomes of the linear and singular goal orientation embedded in *The Island of Dr. Brain*. On a different day, Roger took this same position with another girl playing the game for the first time. The undergraduate working with her was upset by this approach and wrote an unusually critical field note that describes problems with Roger and the game and how they disrupted the 5thD's goals: "She was not learning to solve these problems on her own. He wouldn't tell her the point of the game, or how he was figuring out these solutions, he just commanded her. I guess he saw the goal as getting to the next level, no matter what. It did not matter if she understood or not." Other kids were more amenable to the adult orientation, and after Roger left, the girl and the undergraduate worked together on the problems again, this time reading the instructions and "really" solving the problems together.

The higher priority that the adults we observed gave to process and understanding also translated into a tension between an explicit instruction-driven orientation and a trial-and-error, guessing orientation. The adults tended to stand in for the former orientation and the kids for the latter, particularly when they were not working under close adult supervision. Unless held back by an adult, the kids almost always clicked quickly through the instructions and invariably looked to the hint watch for partial answers before reading the instructions for explicit directions. One undergraduate writes glowingly about a "very smart boy" and notes with impressed surprise how he would "always read the instructions." More typical are the observations of another undergraduate working with a different boy: "When we would get to each new puzzle he did not bother to read all of the directions. I was inclined to read each sentence carefully trying to remember what it said and go through all of the directions before starting each task."

In the most striking example of this tension, one tape shows an undergraduate repeatedly telling a boy that she can't help him because she has not seen the instructions beforehand. As they continue to run into difficulty with the puzzle sequences, she eventually has him read the instructions out loud before proceeding with problem solving. It is not clear, however, that explicit instruction-based problem orientations are the most successful for the game. With the addition of the hint watch, guessing is often a much more expedient solution than following instructions. For example, the two are unable to figure out the tip-o-meter despite reading the instructions out loud. The undergraduate complains to the site director that the instructions are "weird," and the boy eventually solves the problem by using his hint watch charges. In the subsequent rat-driven elevator problem, the boy rejects the instruction-based orientation and solicits help from another kid at site, who shows him how to use the hint watch to get half the answer and then guess at the other half, a strategy that some other kids developed. Although the boy was willing to test out the undergraduate's instruction-driven strategy, eventually the hint-and-guess strategy wins out.

Unlike simulation, strategy, and scenario games, adventure games such as *Dr. Brain* require a narrowly defined set of "correct" inputs in order to proceed through the game. There are clear right and wrong answers, and winning conditions are assessed based only on these answers to a series of problems. The kids expressed satisfaction and even glee at "outwitting" the game, but adults tried to steer them toward solving the problem without guessing. Adults at the club did try to guide children toward understanding the problem rather than guessing or getting an answer from someone else, but it was a struggle for them to initiate the kids' engagement with academic content. Even these moves by the adults were oriented toward the problem's structure rather than its content. For example, they pushed kids toward comprehension of the instructions for solving the puzzle rather than toward understanding, say, how circuitry works or what the difference between a reptile and an amphibian is. Much of this orientation is determined by the fact that puzzle completion affords immediate progression to another puzzle rather than continued engagement with a given content domain. When a puzzle is completed, the computer responds with a plaque of achievement and moves quickly to an entirely different challenge.

Competition, Achievement, and Knowledge Display

Competition is a basic feature of all game-oriented children's software, but games such as *Dr. Brain* that have clear parameters for competition, well-defined obstacles, and unambiguous recognition of success invite the most dogged orientation toward winning, often at the expense of actually mastering or making sense of the content embedded in the game. This peculiarly academic brand of competition as translated to a recreational domain is not immediately consequential for sorting or assessment performed by educational institutions and testing. This logic applies to "the paradoxical concerns of those who are most likely to succeed with altogether inconsequential competitions. It is the story of . . . intense work constructing competition. Continual quizzes, tests, exams, special project, sports events, and so on produce complex ranking that are displayed in plaques, trophies, special citations. Individual qualities become public events" (Varenne and McDermott 1998, 18; see also Goldman and McDermott 1987).

The Island of Dr. Brain is part of this cultural construction and display of a form of competition that is institutionally separate from schooling, but is tied to a related discourse and orientation toward achievement. The game has clear endings to each puzzle, where kids receive a plaque keyed to different levels of difficulty (bronze for novice, silver for intermediate, and gold for advanced). Kids can view these plaques on a progress chart throughout the game, and the game tallies their overall score once they have traveled through the entire island. Game play is punctuated by these small recognitions of achievement, which kids will point out to their partners if they haven't noticed, or, if they are playing alone, they may even call out to others at other parts of the club. Although these "token" achievements don't serve to sort children in school or even in the context of the 5thD, they still matter to the kids. One undergraduate writes: "Dr. Brain kept on giving [the boy] bronze awards. [The boy] kept on saying that this kind of sucked. Why couldn't he get better than bronze, like silver or gold? He wanted better awards." One boy, working alone with the site coordinator's occasional help, was struggling with a particularly difficult problem and finally solved it, apparently by repeatedly guessing. He gleefully shouted, "Yes! I did it!" calling out the site coordinator's name. "I got it! I got it!" he continued to shout, dragging two other kids to the computer to show them. "I did it!" Undergraduates also participated in constructing and displaying these recognitions of success. "Upon the puzzle's completion, I exclaimed to Herbert, 'Excellent job!' The grinning Herbert proudly replied, 'Thanks.'"

Certain puzzles in *Dr. Brain* are specifically designed for recognitions of levels of achievement through "beating" the game. For example, one puzzle is timed and will tell you if you beat the record time for completing it. Kids would usually orient to the stopwatch in the corner and gauge their success on whether they beat the record. In one recorded instance, Chris and an undergraduate mutually orient to this goal as the game sets it up:

C = Chris
UG = Undergraduate

1 C: Well, we got the tree done. Nobody can, nobody beat this yet, know that?
2 UG: Really?
3 C: Nobody's beat it in this kind of times.
4 UG: I know, you're doing it really well.
5 C: Nobody's beat it . . .
6 UG: You got the diving board, and the tree, and what's this . . .
7 C: Nobody's ever gone this fast, you know that, I've probably, I'll probably beat the record.
8 UG: You probably will. I've never seen anyone do it this fast. That's a hard one; you could put that one back for a while.
9 C: I think this actually goes in there.
10 UG: Oh, wait, you're right. Where do you think that one goes?
11 C: Oh, I see something.
12 UG: There you go.
13 C: I'm going really fast.
14 UG: Yep.
15 C: How fast do you think I'm going? Faster than anybody.
16 UG: It's timing you down there.
17 C: It is?
18 UG: Yeah.
19 C: Oh gee.
20 UG: Seven minutes.
21 C: Then I'm beating the record. I'm beating the record.
22 UG: What's the record, do you know?
23 C: I don't know, but I'm probably going to beat the record because it looks like it. I've never gone, nobody's ever gone this fast.
24 UG: OK, where is that thing? What is that thing called, that little like . . .?
25 C: No, that goes right there. Yeah, this goes somewhere.
26 UG: You're almost done.
27 C: I might even beat the record. Where would this go?
28 UG: Right there.
29 C: I couldn't have done it without you, you know that?
30 UG: I didn't do much. You did most of it, and you only have one more piece left.
31 (Chris finishes the puzzle and gets a message that he broke the novice record.)
32 UG: Yeah, you broke it.

Chris is particularly attentive to the time record (lines 3, 5) and his own speed in completing the puzzle (lines 7, 13, 15). The undergraduate encourages this orientation toward beating the record, pointing out the stopwatch in the corner (lines 16, 20) and asking him if he knows what the record is (line 22).

In another instance of play with this same puzzle, an undergraduate and Cathy similarly cemented a common recognition of "beating the record," although they eventually discovered that there wasn't actually much of a record to beat: "Cathy figured out what to do very quickly and was finished with the puzzle in six minutes, which I considered a great achievement. Dr. Brain announced that we had beaten the all-time novice record. I was pretty excited, and so was Cathy, until we found out the all time novice record had been 99.99 minutes. But I told her that she had done really well, and she believed I was sincere about it."

Given that *The Island of Dr. Brain* is a series of puzzles of this sort that culminate in the final "beating" of the entire game sequence, completion of the overall island adventure is greeted with enthusiasm. The game provides a lengthy closing animation, special thanks from Dr. Brain, and an entertaining animation for the final credits. We have only two instances on tape when the game was completed, but both invited enthusiastic responses. In one sequence, a boy completes the game with his dad's help, showing his enthusiasm about finishing the game and receiving recognition from others at the club.

P = Paul
N = Norma
GS = Graduate Student

1   P: (Completes final puzzle, and game screen changes.) Oh, yeah. Cool. (Game cuts to final animation sequence.) Oh yeah! It's a huge boat! (Turning to his dad.) I beat the whole island. (Reads final messages and watches animation.)
    (Dad gets up and leaves.)
2   P: (Continues to watch animation.) Oh cool man. (Points at screen.)
3   N: (Leaning in from neighboring machine.) What is that?
4   P: Yeah, it's an old hidden ship. It's an island that looks like a boat. What the heck are you doing, ship? (Scene cuts to credits sequence, and P reads messages and credits.)
5   GS: (Arrives on camera.) Did your dad go home?
6   P: Nnn-nn. He went to make a phone call.
7   GS: You finished! Congratulations!
8   P: (Continues to watch animation.) Oh yeah! I beat the game! (Raises arms and waves them in the air.)

9   GS: Congratulations! What are you going to do next?
10 P: I don't know.
11 GS: How long did it take you to do it?
12 P: Ohhh, two weeks, maybe three.

All of these examples from *Dr. Brain* rely on game play in relation to clear and stable goals determined by the game, which invites celebration and knowledge display upon achieving those goals. As undergraduates at the 5thD worked to build kids' confidence in their intelligence and abilities, they also reinforced and produced these displays of achievement. Token awards presented by Dr. Brain became social occasions to display achievement. High scores and records were a way of producing success in competition with others in a way that was somewhat removed from one-on-one competition between kids at the club, something that the 5thD philosophy discourages. Unlike a school exam, these game practices are not about sorting children, and they do not explicitly produce failure, but they are about producing a subjectivity as an intelligent and competent person. Within the context of the 5thD, where groups of children mingled with educationally minded adults, the game mediated the contexts of school and play. Adults and the game performed academic achievement in a context that was relaxed, enjoyable, and rich with peer-group interaction, strengthening kid's identification with academic content. At the same time, the game was also in tension with the 5thD philosophy that encourages collaborative and noncompetitive learning processes. These tensions became especially visible in the case of one boy's engagement with the game.

Roger and *The Island of Dr. Brain*

A closer look at Roger and his play with *The Island of Dr. Brain* illustrates some of the achievement-oriented investments and subject formation that the game invites. Roger has already appeared in earlier examples as a game expert who often looked over other kids' shoulders and gave them answers to the puzzles. He had a reputation at the club as academically competent, especially in the area of computer literacy. One undergraduate describes in a field note how he was "really good at the game," and another how he "really impressed me with his knowledge of the game." After his completion of the game, Roger became known at the club as an expert at *The*

Island of Dr. Brain and appeared frequently in subsequent tapes of game play, either helping or heckling other kids. Roger provides a good case for observing a child who orients toward an academic genre of participation. At the same time, his relation to the game became a source of tension at the club. In the sequence given earlier, the undergraduate was upset at Roger's "commanding" attitude toward a girl who was playing. "If she got something wrong, he would tell [her] that she was stupid and did it wrong. If he got it right, he would say that he was the smartest person." On another day, he was working with two other boys on the game. The undergraduate sitting with him wrote: "I did not realize that Roger was completely dominating the game until the two boys got up and left from lack of interest. . . . I like Roger. He is a fun kid, but working with him can make me frustrated due to his limited focus."

Roger's displays of knowledge and achievement created tensions between him and other children as well as in the club organization. He became an object of interventions by the adults at the club. On one occasion, he got into a dispute with the club organizers when he claimed that he had finished the club's maze and should be awarded the title "Young Wizard's Assistant," but the organizers could not find documentation that he had actually completed all the activities. On one day, the site coordinator and site director tried to pull him away from *Dr. Brain* to deal with the issue of his status at the club, but he was busy "helping" another girl play. The undergraduate complains in the field notes for the day that his "idea of helping was to do everything for us. . . . [The site director and coordinator] saw what was happening and told Roger to do something else besides play Cathy's game for her. . . . In exasperation, I covered the computer screen with my hands." Roger eventually left with the site director but then returned, and the undergraduate wrote that he "told us that he could do this puzzle in seven moves. I told him that was nice, but we could do it on our own. He left." These everyday negotiations, displays of expertise, and squabbles at the club were commonplace, and Roger was not a particularly troublesome kid. What was unique about him, however, was his identification with *Dr. Brain* and how that identification became a source of both positive and negative assessments of his activity at the club.

The video record provides more detail on how Roger engaged with the game and others at the club. The first set of examples is from the day Roger

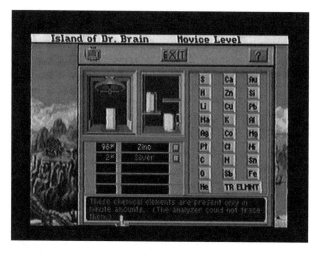

Figure 2.7
Chemical elements puzzle in *The Island of Dr. Brain*. Screen shot reproduced with permission from Vivendi Universal Games.

completes the whole game sequence of *The Island of Dr. Brain* during one club period. He had played bits and pieces of the game previously with other kids, but this is the first day in which he gets sustained time with the game and, with adult help, moves through puzzle after puzzle. In this first example, Roger has just begun to work on a puzzle that involves identifying the chemical code for elements in a set of objects—a tin cup, a zinc bar, and so on (figure 2.7). When the puzzle pops up, he reads the instructions and then tries clicking around to determine the nature of the task. The adults working with him have temporarily been discussing other matters, but he calls them back to the task with a question.

R = Roger
A = Adult
SC = Site Coordinator

1 R: What am I supposed to do? I don't get this.
2 SC: OK, did you analyze it? It says: "These chemical elements . . ."
3 R: (Pulls down another screen of directions and reads, moving pointer over lines.)
4 SC: Oh, you're doing trace elements OK, here.
5 R: Ahhhh! I see. (Starts to read the description of the element he has to find. The object under question is a zinc bar.)

6   A: Oh, do you get the hints?

7   SC: "Blank" oxide (referring to the description, which gives a hint that the answer is a "_oxide").

8   R: Carbon. Blank? Blank?

9   A: See the blank here? (Points to screen.) They're saying fill in the blank.

10   R: Yeah, I know.

11   SC: It's like the sun block people put on their face. . . . You know, people put it on their nose . . .

12   R: Yeah, what is it?

13   SC: What is it called?

14   R: SPF.

15   SC: No. There's a thing that completely blocks it out.

16   R: What? Blank?

17   SC: Zinc oxide, maybe?

18   R: Ziiiinc . . .

19   SC: Have you ever heard of that?

20   R: (Nods.)

21   SC: It's the really white stuff. So you have to find that.

22   R: What's the "Z"? (Points to "Z" in table of elements.)

23   SC: Go up one. That's the zinc. See it up on top?

24   R: (Selects Z for zinc and gets the first element identified correctly.) All right.

25   SC: OK. Now you're doing the next one. It's 2 percent. It says: "These chemicals are present only in minute amounts. The analyzer cannot trace them." So that's the hint you got before, which is the trace element, which means there wasn't enough of them to pick up.

26   R: (Selects "Trace Element" and successfully completes analysis of the first object.) Allll right. Zinc bar. . . . (Places the zinc bar to the side and puts the tin cup in the analyzer.) This is tin. I know it already. Tin . . .

In this sequence of activity, Roger orients quickly to the game's suggested task structure for a new problem: read the directions, determine what the problem is, get the correct answer to the problem, and display knowledge. The game's call to action is: process procedure, execute procedure, solve problem, record solution (fill in the blank). When Roger falters in determining the procedure, he enlists the adults' help: "What am I supposed to do? I don't get this" (line 1). Roger and the site coordinator orient to the instructions, and then the initial recognition occurs, "Ahhh. I see" (line 5), as he is able to decode the instructions and recognize the call for action. Both Roger and the site coordinator then shift their orientation toward the content domain and solving the problem: What kind of oxide is it? (lines 7–10). The site coordinator then tries to get Roger to fill in the answer by providing some hints, though she eventually must give him the answer: zinc (lines 11–17). Roger responds with another act of recognition: "Ziiinc," he says in an extended, low tone and nods to the site coordinator, confirming that he understands the answer (lines 18, 20). He thus positions himself as the subject who has responded to the call for a particular answer. For the remainder of the clip, the site coordinator guides him in locating zinc on the list of elements, and he enters the answer: "All right" (line 24).

This mode of interaction with the puzzles, where Roger decoded the instructions, executed them in solving the puzzle, completed the puzzle, and moved quickly on to the next was typical of his engagement throughout most of the game. As he worked through a puzzle, he punctuated each successfully completed step with an "All right" or an "Ahhh!" of recognition. In this way, he repeatedly enacted the academic genre of participation, taking the stance of someone whose knowledge and competence is being tested and assessed. Roger's brief utterances of recognition were subtle but repeated frequently. On other occasions, he made more explicit statements that pointed to his increasing subjectification in the terms of academic achievement as suggested by the game.

In the following clip, from the same day as the previous clip, Roger has just completed the Tower of Hanoi problem in *The Island of Dr. Brain*. His adult partner has been engaged with another child while Roger works on the puzzle, and he tries to draw her attention to the fact that he has solved the puzzle, self-identifying himself as a "smooth" problem solver, displaying his competence.

R = Roger
A = Adult

1   R: Now I have it solved! (Turns to adult who is still preoccupied with other child.) I got it solved . . .
2   Computer: Congratulations! You've earned a bronze logic-sequence prize!
3   R: (Still trying to get adult's attention.) I did it . . .
4   A: (Turns back to Roger.) Sweet.
5   R: I'm smoooooth.
6   A: How many moves did you do it in?
7   R: Seventeen.

As described earlier in this chapter, the singular and linear goal orientation of *The Island of Dr. Brain* encourages kids to input correct answers, often

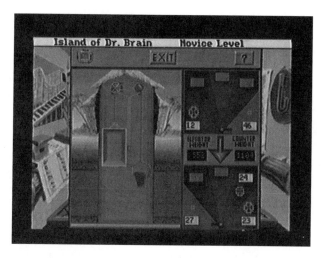

Figure 2.8
Rat-driven elevator problem in *The Island of Dr. Brain*. Screen shot reproduced with permission from Vivendi Universal Games.

at the expense of engaging with educational content. Roger was a particularly adept strategist in this regard. In many instances of his play, I marveled at how he was able to identify quickly the minimal conditions for solving a task, delegating as much problem-solving effort as possible to other helpers in the neighborhood and getting through the problem. The next example is from one of the first instances of Roger's exposure to *Dr. Brain*. He is working with another kid, Herbert, and they are just beginning "the rat-driven elevator" problem (figure 2.8). The site coordinator occasionally checks in on their play. This is the first time that either of them has encountered this problem, and they are exploring and trying to decode the problem space. The task is a complex one. They are asked to determine how many spokes, on two different gears, are required to balance a counterweight with the weight of the elevator. They spend quite some time keying in different answers and trying to figure out the nature of the problem, enlisting the site coordinator's help. They try various solutions, but the elevator continues to fly into either the ceiling or the floor, toppling the crash-test dummy inside. They eventually begin to enjoy simply watching the dummy crash time after time, moving from achievement orientation to pleasure in this spectacle. After almost ten minutes in which they continue their trial-and-error tactics, they finally happen on the correct answer.

R = Roger
H = Herbert
SC = Site Coordinator

1   R: OK, fifty-six, fifty-one, and seventeen. You have seventeen and forty-eight.
          Forty-eight. OK, let's try it.
2   H: Yeah.
3   R: I love doing this.
4   H: Yeah, this is it. Yep. Nope. Nope.
          (Elevator crashes.)
5   R: Ahhh!! I love that.
6   H: It must, it must be fifty-one. Oh, man.
7   R: This is so hard.
8   H: Eighteen teeth. Watch this, watch this, watch this.
9   R: You think this is right? No, he got tired. (Elevator crashes.) Ahh!!! (Laughs.)
          I love this!!
10   H: Eight, twenty-one. Nooo!!!!
          (Elevator crashes.)
11   R: I love doing this.
12   H: Thirteen. Yeah. (Elevator is lowered successfully.) Oh my gosh. We got it.
          We got it!!! Yeah, [site coordinator's name]. We got it.
13   SC: All right!
14   R: And we did it by guessing, too!
15   H: I know, huh!
16   R: We're so good. Yeah, we can ride it.
17   H: Yeah.

This clip records a gleeful moment, with Herbert calling out to the site coordinator about their accomplishment and the two boys mutually congratulating themselves (lines 12–16). Even though they are still proceeding along the game's sequential logic, they manage to claim a small space of achievement for themselves, which is not tied to the procedure for action as suggested by the game's explicit educational goals. They are still, provisionally, heeding the call to action: working on decoding the instructions and getting the correct answer. Most important, they persevere and achieve "mastery" in terms of the game's measures of success. They are particularly happy at having "tricked the system" by getting the right answer through guessing. Far from detracting from their sense of mastery, this accomplishment serves as a display of achievement. "We're so good. Yeah, we can ride it" (line 16). Our tapes show Roger revisiting the same problem on a subsequent day and mobilizing the guessing tactic that he developed with Herbert—thus abandoning any attempts to decode the nature of the problem and reproducing the guessing heuristic in another context.

R = Roger
A = Adult

1 R: Now I can open it. (Opens door to rat-driven elevator problem.) I LOVE this puzzle! This is so funny. We just guess. Me and my friend did it, and we just kept guessing.

2 A: Oh really? It kept kicking me out.

3 R: Watch this.

4 A: So what you do is (pointing to screen) you divide the elevator weight into the counterweight.

5 R: (Inputs a solution, and the elevator crashes.) Oh no!

6 A: Crash-test dummy!
(Game states correct answer and then resets puzzle.)

7 R: Oh! It was twelve and twenty-four. Oh, I see. (Game has kicked him out of the puzzle, and he reenters it.)

8 A: How many times does 428 go into 1,284?

9 R: I have no idea. I'm just guessing. It works. (He eventually does it.) Yeah! (Elevator is lowered successfully.) Am I good or what?

10 A: Pure luck.

In this clip, Roger begins by announcing his guessing tactic, but the adult repeatedly tries to get him to orient to the procedure as suggested by the game and to the subject position of problem solver: "So what you do is you divide the elevator weight into the counterweight" (line 4). "How many times does 428 go into 1,284?" (line 8). This disjuncture between the narrow definition of achievement called forth by the testing situation (input the correct answer) and the more demanding definition of achievement called forth by the adult (follow the correct procedure before inputting the answer) points to a gap between the formal measure of achievement and the process that is meant to underlie the measure. Roger's ability to exploit this space between a formal recognition of mastery and an actual expenditure of personal effort is similar to other test-taking tactics developed by kids who learn to "play the system." In doing a later puzzle, Roger guessed correctly again and looked smugly at his partner. "I did it. And I just guessed. I'm so good at guessing."

A few months later Roger was well established as the site expert on *The Island of Dr. Brain*. I have already described field notes with frequent references to his checking in on other kids' play of the game, often displaying his knowledge and giving answers. In the following clip, Roger's help is being actively solicited by a younger child, an eight-year-old Chris, also previously mentioned in this chapter. During the course of the day's play, Roger made frequent appearances over Chris's shoulder, giving him instructions on how to play. Chris would often call out to Roger to come help him if he was stuck. This brief clip illustrates a typical sequence of interaction, where Chris encounters a puzzle, Roger identifies himself as an expert, and Chris solicits help.

C = Chris
UG = Undergraduate
R = Roger

1   C: Now we are here again.
2   UG: Oh that's hard.
3   R: This is hard. I'm pretty good at this, though.
4   C: OK, I need some help.
5   R: All right.

This instance was not an isolated one. Throughout the course of his play with Chris, Roger repeatedly made statements such as "I'm very good at this" or "I'm really good at this" with regard to almost every puzzle that appeared in the game. Other children also displayed achievement when completing a task, but Roger displayed a persistent subjectivity as an expert at this game across instances of his own play as well as in participation with others' play. This subjectivity is distinct from activity-specific acknowledgments such as "I did it!" that punctuate ongoing game play.

Roger's sense of mastery of the game is documented in the video record through repeated occasions of engagement with *Dr. Brain*, leading to his increasingly public and open displays of game mastery to other kids and adults at the club. Both the game and others repeatedly recognized Roger as a particular kind of learner, and he repeatedly responded to this "hail" (Althusser 1969) from the game and others and subjectified himself to this shared sociotechnical formation. Insofar as Roger had become identified and self-identified as an expert at this game within the club, his mode of engagement stood out from that of other children. Yet as seen in the other instances of children engaging with and learning from *Dr. Brain*, his orientation to beating the game was one that other children quickly modeled. Despite adults' efforts to redirect the kids' engagement toward more content-oriented and less goal-directed models of play, the academic genre embedded in the game and the ready availability of academic genres of participation tended to override the adults' ongoing interference patterns and reform-minded orientation.

Conclusions

The case of Roger's engagement with *Dr. Brain* illustrates processes of negotiation and alignment between media genres and participation genres that are resources in kids' everyday lives. Although some adults in the 5thD have made an effort to introduce other genres of participation to the repertoire of practices surrounding *Dr. Brain*, the academic genre is resilient because of the force of the game design and the ready availability of elements of the academic genre of participation in the culture of childhood. Across the circuit of culture, Roger shakes hands with game designers who reinforce particular genre recognitions, despite local adults' differing educational philosophies.

Although adults at the club do work to temper the competitive edge of the academic genre of participation, they also reinforce it through acknowledgment of academic accomplishment. Knowledge displays are oriented to both adults and other children, but adults form the most appreciative audience, structuring the reception context with their recognition of success. "Congratulations!" "I've never seen anyone do it so fast." Through the repeated performance of this kind of commonplace interaction, the software title delivers on a certain set of promises made in the subtext of the genre's marketing. As the *JumpStart* ad copy suggests, "When kids succeed, they feel confident. When they feel confident, they succeed." Interest in and engagement with content is secondary to the attribution of success, positioning a child in a status hierarchy defined by "winning" in a competitive exercise marked as academically significant. Although latent within the 5thD setting, the academic genre is readily available as a resource for both kids and adults to mobilize, particularly through the reinforcement of a software title in the genre. The particularities of the software design are important to keep in mind, as are the particularities of the 5thD setting that inflect the software's use. The collusion between very specific game features such as score-keeping functions and ways of getting hints encourage certain forms of engagement. One can imagine a game that has packaged similar content in slightly different ways, manifesting substantially different outcomes in practice.

Although the cases of engagement with *Dr. Brain* I have described may not represent ideal kinds of learning outcomes for a setting such as the 5thD, they do represent successful genre recognitions by children that tie together sites of consumption and production: players recognize themselves in the genre suggested by the designers and marketed by the advertisers. Just as the 5thD's reform-oriented educators struggle on a daily basis to introduce new kinds of participation genres into the children's already densely structured lives, software designers have struggled to introduce and promulgate new idioms of engagement through innovation in design. Other software titles do more to complicate the distinctions between entertainment and education that dominate children's lives, but the academic genre tends to reinscribe existing participation genres that are dominant in schooling and academic achievement. When presented with a title like *Dr. Brain,* kids are quick to recognize an academically oriented genre of media and participation because this genre is reinforced in settings and institutions distributed across the circuit of culture.

Children engaging with *Dr. Brain* participate in a genre of academic preparatory play that Hervé Varenne, Shelley Goldman, and Ray McDermott (1998) frame as typical in upper-middle-class schools. These authors describe a question-and-answer game called *Screw Thy Neighbor* played at an elite middle school, where kids cover school topics within a quiz-show genre. They conclude that compared to standardized testing, " 'Screw Thy Neighbor,' like the Balinese cockfight, does not do anything. Functionally, it is 'just' deep play, and culturally, it is the stuff of life" (1998, 112–113). As one teacher explained, "You make it into a contest, and suddenly everyone wants to be an expert at defining vocabulary words" (114). Varenne and his colleagues comment, "In their vocabulary, [the students] say simply that the competitions of everyday life are 'fun.' Competition transforms the boring into the interesting" (114). Games such as *Screw Thy Neighbor* and *Dr. Brain* are preparatory for "real" school tests and productive of a competitive, success-oriented structure of participation.

As one of its sociocultural effects, software in the academic genre can produce a certain form of class identity, merging the traditional status of academic content with a highbrow media form, packaged in an entertainment-oriented visual vernacular that can travel across the school boundary into the home and other recreational contexts. Thwarting early developers' intentions, software in this genre has become another site for addressing achievement anxiety in parents and for instilling the stance of upwardly mobile achievement in children who seem to have been born into success. Reform efforts that rely on educational media must produce innovative content as well as innovations in distribution mechanisms and contexts of play in order to have a systemic impact. In this chapter, I have stressed the conservative tendencies of the academic genre.

Activity in the 5thD demonstrates, however, that structuring sites of reception can substantively alter these conditions of social reproduction by downplaying the competitive dimensions of play with this kind of software. The history of the academic genre's emergence also demonstrates that learning software can be oriented otherwise. The design and marketing of games does matter and can translate to substantive differences in forms of engagement. In fact, just as the founders of the industry criticize current products, many of the educators at the 5thD resisted the newer titles through the 1990s and continued to support the Apple II and the early products produced for that platform. These products are entering technological obsolescence, however, as CD-ROMs and DVDs have thoroughly replaced floppy disks, and Apple IIs are becoming difficult to maintain for even the most avid hobbyist. Just as PC users have expanded from a hobbyist, small-scale community to a mass market, learning software has also become a mainstream commodity, enlisting a different set of interests and actors as it shape-shifted from an educational reform effort to an established market niche. Richer multimedia formats rely on visual vernaculars that require a certain level of production value. For a niche market such as edutainment, limited development budgets have come to be allocated primarily to maintaining production value and reusing assets rather than to producing social and cultural innovation.

School-based content, achievement concerns, a culture of competition, and children's visual culture are the cultural forms that get packaged into the durable disposition of software in the academic genre. The commodification of learning has unique dynamics, mobilizing highbrow academic goals and failure anxiety as primary marketing tropes that find resonance in a portion of the middle class that has become the consumer market for edutainment. In the context of the 5thD, my fellow researchers and I observed the enactment of some of these logics of competition and achievement, but also the efforts to create a setting where learning is not primarily an occasion for sorting kids according to academic achievement. In this chapter, I have described "learning through play" as a political, economic, and technological event, produced across a heterogeneous but linked set of social contexts. The next chapter takes a similar approach to the production of "entertainment" as a genre structurally opposed to that of academics and with deeper roots in children's commercial popular culture.

3 Entertainment

In 1980, while Ann Piestrup (McCormick) was developing her first software titles for the Apple II, Gary and Douglas Carlston were beginning a new venture from their apartment in Eugene, Oregon, shipping plastic sandwich bags with floppy disks that they had copied by hand. This home-brewed company eventually became Brøderbund, the publisher of titles such as *Just Grandma and Me* and *Myst* that defined CD-ROM-based multimedia in the mid-1990s. In 1982, Brøderbund bought the rights to publish *Bank Street Writer*, a word-processing program for children developed by educators at the Bank Street College of Education. This initial foray into children's software laid the groundwork for Brøderbund to become one of the leaders in children's CD-ROM publishing, producing titles such as *Where in the World Is Carmen San Diego* and the Living Books series.

In a 1999 phone interview, Gary Carlston described the company's orientation to children's software, which combines an appreciation for childhood curiosity and wonder with a primary orientation toward children's pleasures rather than academic achievement:

None of us had degrees in education. We didn't want to go out and make all these pedagogical claims. On the other hand, we had developed some of these products for a reason. . . . Basically, our idea was to do products that we ourselves found interesting. That's not a real narrow niche. That's just kind of whatever seemed fun. And even things like *Where in the World Is Carmen San Diego* weren't done for the educators. They were done because of our own childhood fascination with almanacs.

Where in the World Is Carmen San Diego, released in 1985, represented Brøderbund's first major commercial success in the edutainment market. By the mid-1980s, the console video market had crashed, as had the fortunes of Atari, and Brøderbund's investors were looking for another niche. Carlston continued:

Our investors wanted us to move away from that uncertain world and find a niche that we could be the leader in. And so we started to do fewer and fewer games, just straight adult or not adult, but typical computer games and to try to get more and more and more into the education area, where it was thought that because of our background in games, that we did more exciting educational products than the more boring educators did (laugh). We had a lot more graphic technology at our disposal at that time because we had a little more experience with that.

In contrast to TLC and Davidson & Associates, Brøderbund was a video game company that expanded into the children's software market. It and companies such as Voyager made their name with multimedia and CD-ROMs rather than with classic educational titles and genres established on the Apple II platforms, and they were most well known for ushering in a new era of graphical sophistication for interactive media. More (self-described) hippie than missionary, Gary Carlston gives voice to the enlightened but boyish curiosity, humor, and antiauthoritarianism that came to define a genre of children's software that focused more on entertainment than on academic achievement. This chapter locates the efforts of software developers such as Carlston in the history and context of the children's software industry and then describes how related elements of visual culture manifested in the play at the 5thD.

Childhood Pleasure and Play

Unlike software in the academic genre, many of the titles produced for children in the 1990s and later owe their inspiration more to television and video game culture than to the classroom culture. The spread of CD-ROMs and multimedia computers marked the entry of children's software into the logic of visual culture and idioms developed by movies, television, and video games. Gary Carlston's celebration of childhood amusement and pleasure is indicative of an orientation that identifies more with the child than with the educator, with indulgence more than with achievement. Unlike what you see in many adult-oriented video games, the content in children's software and games is wholesome and family-friendly, even in the entertainment genre. Entertainment idioms generally appear as *one element* of titles aspiring to educational or developmental goals. For example, *Math Blaster* incorporates a space fantasy and shooting scenes to motivate engagement with math problems. In *Dr. Brain,* an island adventure forms the visual and narrative setting for a set of cognitive tasks. It is thus impossible to isolate elements of hedonistic play and visual culture in particular titles and companies. The multilayered malleability of interactive multimedia means that multiple genres can coexist within a given software package. In children's software, this malleability has resulted in titles that mix the conflicting entertainment and academic genres.

A child-centered popular culture has been growing in momentum ever since the establishment of children's fiction and comic books, and it has expanded into more and more genres of toys and mass media. The television was a turning point in creating a direct marketing channel between cultural producers and children. In his history of toys, Gary Cross describes how through the 1940s, most toys were still advertised and sold to parents and pitched as a tool for parent-child bonding, picturing fathers and sons together around a train set or daughters mimicking their mothers with miniature cooking sets and baby dolls. In the 1950s, this approach began to change. "Television took the toy beyond the worlds of parent, trained sales people, educational experts, and, most of all, tradition. The new medium made possible a constantly changing culture of play that appealed directly to the imaginations of children" (1997, 162). The 1980s saw the growth of toys and characters that often offended parental sensibilities, creating violent fantasy worlds for boys and sexually mature, glamorous consumer role models for girls in characters such as Barbie. Cross concludes that "[o]nce children are old enough to enter the world of peer culture and consumerism, educational toys have relatively little influence" (233).

Much as Thomas Frank (1997) describes in *The Conquest of Cool*, the children's counterculture, like the counterculture of the 1960s and 1970s, has been taken up as a powerful advertising trope, marketing freedom from everyday discipline through consumption. In describing the history of children's media, Sarah Banet-Weiser notes that "the 1950s also mark a shift in authority figures for adolescents; whereas in previous years, family, church, and neighborhood provided advice and guidance about how to construct one's identity, with the rise of the teenage market the commercial mass media came to powerfully stand in as authority figures for children" (2007, 25–26). Steve Kline sees the development of the children's market as similar to the development of the youth market, but with its own unique challenges. "The points of contact between the marketplace and children were few and unsuited to advertising. So it fell to television to open up new lines of communication with children, making marketing to young children possible" (1993, 165). He describes the Mickey Mouse Club as a turning point in the development of a distinct children's consumer culture because it focused on a children's subculture formed by television (166–167). Another turning point was Mattel's marketing of the Burp Gun in 1955 through the Mickey Mouse Club.

> Children obviously couldn't afford to buy the Burp Guns themselves, the toys were too expensive. They had to convince their parents to buy them and to sympathize with the urgencies of faddish children's products. The Burp Gun was hard to justify on the grounds of enhancing education and skills. Children had effectively to lessen their parents' concern to have only educational toys and convey how important the special nature of this object was to them. The success of this television promotion was enough to convince some toy marketers that it was possible to change the family dynamics around consumption by TV campaigns directed at children. (Kline 1993, 168)

Television created a link between adult media producers and children that has greatly expanded children's cultural resources and defined a new cultural domain, an expansion that has gone hand in hand with a rapid increase in spending on children. This child-centered cultural production was initially attentive to parental concerns and yet began to nurture an emerging antiauthoritarian youth culture that pushed back at parents' protective and achievement-oriented values.

Two aspects of this child-centered orientation are resistance to adult achievement and progress goals and the definition of children's culture and pleasures as ecstatic and based in instant gratification. In *Sold Separately*, Ellen Seiter (1995) describes how resistance to adult culture goes hand in hand with hedonistic fantasy. She contrasts the educational or developmental orientation of toy ads in *Parenting* magazine with the ecstatic and utopian world of commercials aimed at children: "A separate children's playground culture and street culture has existed as long as children have lived in cities and gone to school: but this was a culture produced by children and passed on from child to child. A similar kind of culture is now produced by adults and offered to children through the mass media from toddlerhood onward" (117). She comments on how "[c]ommercials seek to establish children's snacks and toys as belonging to a public children's culture, by either removing them from the adult-dominated sphere or presenting these products as at odds with that world" (117). Commercials present an ecstatic and pleasure-seeking childhood that resists parents and teachers' cultural norms. Outlining the cultural logic of television ads for products targeted toward children, Seiter writes, "Anti-authoritarianism is translated into images of buffoonish fathers and ridiculed, humiliated teachers. The sense of family democracy is translated into a world where kids rule, where peer culture is all" (117–118). This media-enhanced peer culture has created a new rupture in the family around issues of entertainment and consumption, giving power and voice to elements of childhood that often sit uneasily with adults.

In his discussion of discourses about children's play, Brian Sutton-Smith suggests that the dominant discourse among adults is one of play as progress or play as fulfilling developmental and learning goals. He sees children as exhibiting a quite different orientation, with play used often as a form of resistance to adult culture and displaying a fascination with irrational fantasy that he calls *phantasmagoria*, characterized by pain, gore, sexuality, and violence that adults continuously work to suppress. "It seems that the history of the imagination in childhood is a history of ever greater suppression and rationalization of the irrational. Paradoxically children, who are supposed to be the players among us, are allowed much less freedom for irrational, wild, dark, or deep play in Western culture than are adults, who are thought not to play at all. Studies of child fantasy are largely about the control, domestication, and direction of childhood" (1997, 152). He suggests that rather than uniformly repress these more phantasmagorical narratives—the nonsensical, irrational, satirical, and violent workings of the childhood imagination—we need to pay more attention to them. In a move resonating with Freud, Sutton-Smith seeks to acknowledge and legitimize these repressed psychological fascinations of childhood play. As adults have sought to ignore and repress the sexual dimensions of childhood relationships, they have also repressed the violent, dark, and irrational dimensions of play in children. He sees contemporary children's culture as giving voice to these repressed elements of childhood.

Borrowing from Michel Foucault (1978), we might consider adult efforts to manage children's play not only as a "repressive" regime that works to silence these dark fantasies, but also as an "incitement to discourse" that gives voice and form to "unnatural" and regressive play in opposition to "natural," wholesome, and productive play. Discourses of childhood play produce not only the vision of "pure" and innocent play, but also a growing corpus of conversations about the regulation, pathologization, and commercial potential of illicit and violent forms of play. This duality is perhaps most evident in the public debates over violence in children's media. Producers of action entertainment argue for catharsis, claiming that they are giving voice to repressed and primal aspects of human nature, whereas concerned parents and activists argue that violent media corrupt an inherently innocent childhood with the pathologies of adult society. Entertainment industries have found an ally in the recognition of childhood agency associated with progressive parenting and the move to less repressive and more pleasure-oriented approaches to child rearing. Despite often polarized differences, all of these positions are co-constitutive of a growing cultural attention to childhood pleasure and imagination. They are contestations over "the natural" in childhood play and pleasure, a discursive space that has expanded along with children's entertainment industries. Ironically, an ever larger discursive, technological, and capitalist apparatus is producing the "discovery" of "natural" and authentic children's play and imagination.

Beginning with comic books and culminating in video games, lowbrow and peer-focused children's culture has been framed as visually rather than textually oriented, relying on fast-paced fantasy and spectacle over realism, subtlety, and reflection. Popular culture has created and given visual form to a wide range of childhood fantasies, including much of the violent and grotesque content of Sutton-Smith's phantasmagoria. Criticism of "trashy" children's media has been a persistent companion to this growth in children's visual culture. With the advent of CD-ROMs and high-quality graphics on PCs, children's software made its entry into this world of contemporary TV-centered visual culture, borrowing from longstanding traditions of character animation and from the interactive spectacles developed in video games since the early 1980s. Contemporary children's popular culture, whether embedded in books, movies, television, or games, is increasingly informed by the logic of visual culture and the high-tech spectacle, the reliance on fast action and high production-value images, music, and sound to create media detached from the mundane experiences of everyday life. Budgets for children's software are being allocated to create more and more visually stunning products that immerse children in compelling virtual fantasies worlds that reference other domains of popular culture more often than the world of their everyday lives.

Visual culture is a mechanism for enlisting children into a peer culture defined in opposition to adults and school, allied with a growing apparatus of commercial media production (Banet-Weiser 2007; Buckingham 1993; Seiter 1995). Visual culture is also a site of intertextual enlistment and translation as different media genres reference each other in a dense web of visual signification (Kinder 1991). Pleasure and fun—whether for adults, youth, or children—are symbolically set off from the instrumental domains of work, discipline, and achievement, highlighting the cultural opposition between "active" production and "passive" consumption. Media industries capitalize on the discursive regime that produces play as a site of authentic childhood agency, specifically by mobilizing phantasmagoria as a site of regressive, illicit, and oppositional power. In children's engagement with spectacle (Debord 1995), there is a level at which it is "just" entertainment, or myopic and inconsequential engagement with visual forms. At another level, however, the engagement is a politically, socially, and culturally productive act that is about participation in certain regimes of identity.

Banet-Weiser argues that children's media such as that provided by Nickelodeon constructs a kind of "consumer citizenship," where "the imagined national community references the community of consumers, united by 'shared codes' of consumption behavior" (2007, 10). This is not simply a matter of giving voice to children's inner fantasies, but of creating a new network of relationships based on media technologies, industry networks, and discourses of childhood—a celebration of childhood imagination in the hands of commerce as much as children. Although children's participation in popular visual culture is active and mobilized, it is, in Louis Althusser's (1969) terms, also part of being interpellated into a subjectivity that is ideological and inseparable from political economic relations. The content of childhood fantasy in children's software can range from the wholesome and the innocent to the phantasmagoric, often incorporating this range within a single product. Developers such as Douglas and Gary Carlston defined the early years of children's multimedia as whimsical, wholesome, and still parent-friendly, more Mickey Mouse Club than Garbage Pail Kids. Yet the logic of the established children's entertainment and toy industry was soon to make its way to this new media platform.

Children's Software and the Dawn of Multimedia

The late 1980s and early 1990s saw the dawn of multimedia, enabled by the spread of PCs in the home and the advent of CD-ROM technology.

When CD-ROMs were introduced during this period, it was the first time consumers could have easy access to high-resolution graphic, animation, and sounds on their PCs. Until the late 1980s, Apple IIs and MS-DOS computers provided the platforms for educational software. After the release of Microsoft Windows in 1983 and of the Macintosh in 1984, the tide began to turn toward more graphically intensive personal computing, and the late 1980s and early 1990s saw the emergence of the new buzzword *multimedia*.

In 1989, the *Visual Almanac,* a product of the Apple Multimedia Lab, was introduced at the MacWorld trade show as a limited-release product to be donated to educators. Using videodisc, a Macintosh, and Hypercard, the *Visual Almanac* heralded a new era of multimedia children's software that would soon shift from videodisc to CD-ROM. Tying together video's graphical capabilities and the PC's interactive qualities, the *Visual Almanac* was the first demonstration of the polished graphical quality in children's software that we have come to associate with CD-ROMs. Voyager was the company best known for making the transition from videodisc to CD-ROM, publishing the first commercial CD-ROM in 1989 and going on to publish children's titles derived from the *Visual Almanac.*

From its inception, multimedia was seen within a lineage of visual culture extending from television and video games. It united lowbrow cultural content with the highbrow promise of the PC. In the early years, the PC's interactive qualities and the promise of active and engaged learning were central in the minds of early educational innovators. By contrast, with multimedia platforms, the focus was on mobilizing visual popular culture to capture children's attentions, tying together education and the styles and genres of contemporary commercial entertainment. Sueann Ambron, founder of the Apple Multimedia Lab, wrote in the late 1980s about the promise of multimedia in education, describing the potential to tap the TV generation's attention.

Multimedia is important in education because it holds great promise for improving the *quality* of education. People have been dreaming about easy access to information that has the richness of multiple images and sounds, and multimedia begins to deliver on the dream.

Students who have difficulty expressing their ideas in writing can now have a new way to communicate and a new class of material to learn from. Children who are used to watching television, listening to music, and playing computer games find multimedia a more compelling learning tool than book-and-chalkboard educational media of their parents' generation. Finally, multimedia allows the user to be an active learner, controlling access to and manipulating vast quantities of information. (1989, 9)

Like TLC, the Apple Multimedia Lab drew heavily from educational research. The developers were educational reformers who sought to enrich children's learning in a way that departed from the usual classroom idioms. Children's "natural" affinity to new technology and visual culture became a tool toward this end. In a 1999 interview, Margo Nanny, a former teacher who helped develop the *Visual Almanac* and was at the Apple Multimedia Lab from 1989 to 1991, described the early years of children's multimedia. Developers shared a research orientation and had close ties to the educational research community.

It was a really fun group. And the thinking was so rich and deep. . . . The people that would come and visit us were the TERCs [Technical Education Research Centers, a nonprofit founded in 1965 and producer of the *Zoombinis* software titles] and the EDCs [Education Development Centers] and all the people doing the most interesting stuff, including Apple Advanced Development. But now, there is nothing like that. There's no place that you can go for those rich conversations about what educators would do with interactive images if they had them.

In a 1998 interview, Bob Mohl, a graduate of the MIT Media Lab and a former member of the Apple Multimedia Lab, explained this liminal position between commercialism and research: "The Apple Multimedia Lab was a very unusual situation. Here you have a big company that's funding all of this stuff, but it's not bringing in really serious company accountabilities and deadlines. You can be really into the process."

Mohl and Nanny went on to adapt some of the content of the *Visual Almanac* for CD-ROM, producing two titles, *Countdown* (figure 3.1) and *Planetary Taxi* (figure 3.2), published by Voyager in the early 1990s, among the first in the new genre of CD-ROM-based children's multimedia. At this point, multimedia was still a garage-shop production. Nanny explained: "*Countdown* was me and Bob and fifty thousand dollars in our garage, and the programmer lived in a tent in my backyard." *Silly Noisy House*, developed by Peggy Weil, was another early children's title published by Voyager, featuring an interactive dollhouse, populated by songs, nursery rhymes, tongue twisters, sound effects, and bears (figure 3.3). Like Mohl and Nanny, Weil was motivated by a holistic approach to learning, mobilizing new

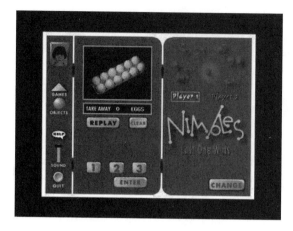

Figure 3.1
Screen shot from *Countdown*. Reproduced with permission from Aurora Media.

Figure 3.2
Screen shot from *Planetary Taxi*. Reproduced with permission from Aurora Media.

Figure 3.3
Screen shot from *Silly Noisy House*. Reproduced with permission from Peggy Weil.

interactive multimedia to stimulate childhood engagement and imagination. In a 2008 interview, Weil described a confrontation she had at an expo where she was exhibiting her CD-ROM:

Someone said to me, "But why should I buy this for my kids when that other project has letters and number drills?" I snapped back, "When you buy a dollhouse for your child, do you insist that it have letter and number drills?" Per the "whole language" philosophy of education, discovering and parroting a tongue twister attached to the pickles in the refrigerator is more valuable and far more educational than phonemes, out of context, plodding down a conveyor belt in *Reader Rabbit*. I feel that singing along to rhymes, especially after having found them, fosters more love of literature, as well as memory retention, than drills.

Mohl, Nanny, and Weil were part of the construction of a new genre of software that was defined in opposition to the earlier, more school-centered genre of edutainment. Though relying on a small budget and a small team, the developers of these titles were able to set the stage for the use of entertainment-quality graphics, music, and sound effects for children's software. Developing graphical expertise for the computational medium was a process of enlisting talent and technique from other genres through the 1980s and 1990s. Collette Michaud, a graphical artist who worked at TLC before moving on to manage the LucasArts Entertainment Company art department, described in a 1999 interview how in that decade artists were drawn from other media forms because very few artists had been trained to produce computer graphics. In 1991, for example,

[c]omputer Games had just shifted over from EGA-16 colors on screen, to VGA-256 colors on the screen. The resolution of 325/600 was the same—pretty low. But having the extra colors opened up a lot of new possibilities for game graphics. Suddenly, we started to attract more artists, but it was still tough to find any who were interested in working on the computer in such low resolution with so few colors.

Most of the artists I hired came from the comic book industry. That was the best place to find artists because they were extremely versatile. They could draw and animate without looking at any reference. I couldn't find artists who knew how to use the computer because there just weren't any. So I tried to find artists who had good drawing skills. That was the most important thing. You could always train them on the tools of the computer.

Around the same time—a time when it was still uncertain whether CDs would become a viable form of computational media—Brøderbund was releasing its first Living Books, such as *Just Grandma and Me* (figure 3.4), and a CD-ROM version of *Where in the World Is Carmen San Diego*. Voyager was another pioneer in the market, making the transition from videodisc to CD-ROM in 1988 with the release of *Beethoven's Ninth Symphony*, the first consumer CD-ROM. In 1991, it released its first children's titles. Brøderbund and Voyager's gamble did pay off, and the early 1990s marked the beginning of the mainstreaming of multimedia computing. In the late 1980s, the *New York Times* began carrying regular reviews of children's software. In one of these reviews from 1992, Peter H. Lewis introduced the term *multimedia* in the mainstream media: "Multimedia is a nebulous thing. Basically it involves adding sounds, voices, animation, video and other eye-catching data types to the simple text and graphics familiar to most computer users" (1992, 2). He explained that to get animation and CD-quality sound, however, a family would need to purchase a CD-ROM drive, at the time still priced at an inaccessible $500. Although it took until the mid-1990s for the CD-ROM to become a standard feature of PCs, even in the early 1990s there was a growing shift toward graphical design as a central component of software production.

Figure 3.4

Screen shot from *Just Grandma and Me*. Reproduced with permission from The Learning Company.

In my discussions with game developers, many commented on the growth of graphic arts budgets through the 1990s. Michaud described how when she started at LucasArts in 1991, there were fourteen artists in the art department, but when she left in 1996, there were sixty-five. Compared to the case in the early years, where programming was the primary trade of computer-game development, graphic artists have taken an increasingly central role. Today, artists will generally outnumber programmers in a game-development team. Michaud explained how artists gradually went from being "just the wrists" to having a stronger and stronger voice in the design of the products as graphical content became centrally defining. For children's titles that are released as part of a series of products, corporations often reuse the same underlying technology and settings, and plug in different graphics, sounds, and story lines to produce a new title. Even in a big-budget game such as *SimCity*, which is currently in its fourth incarnation, you see the same concept being re-released with more sophisticated graphics and produced by larger and larger development teams. The original *SimCity* was the work of a lone programmer, whereas *SimCity 3000* required dozens.

In contrast to the companies incubated by educators as nonprofit ventures in the late 1970s, the multimedia ventures launched in the late 1980s were generally started by technology and media companies. By that time, PC-based gaming had expanded enough to be an appealing new market. Michaud commented on some of the differences in orientation between children's software production at TLC and LucasArts, where she was developing children's titles within an established video-gaming company.

At The Learning Company it was all very politically correct. When I created a character, it was put through rigorous executive meetings, not focus groups. Back then, we never really tested anything with kids. Instead, the president and his executive staff would get involved and say, "Well, that character can't be white, and it can't be blond, and it can't be red, and it can't be a boy, and it can't be a girl, because, if it is, we'll be offending all of these groups." So I was forced to create homogeneous characters. Needless to say, this was a bit tiresome, so I was ready to move on to an entertainment company where it was okay to be controversial. I could create a witch and actually make it look ugly or create a female character and make her look sexy without anyone getting uptight about it.

Whereas the Apple II's more educationally oriented and minimalist platform gave birth to TLC and Davidson & Associates, founded by former teachers, the 1990s saw a shift toward an entertainment orientation in children's products. Companies such as Microsoft and Apple were incubating their own ventures into children's software that would have a strong research and educational orientation, but would take into account the more graphically intensive and entertaining formats being developed for game consoles and arcades. As a commercial market, these new ventures were not under the same constraints as classroom software, and their creators were given more freedom to develop content that appealed directly to children. The shift was from a pedagogical perspective that sought to educate children to an entertainment orientation that sought to give voice and shape to children's pleasures. Gaming companies such as Brøderbund were beginning to see children's software as an area where they could create graphically exciting, entertaining, *and* family-friendly products. Maxis's *SimCity* became a hit product that spanned the entertainment and education markets, although it was not originally intended as an educational title. Edutainment was an expanding site of negotiation and struggle among educators, entertainers, programmers, artists, and businesspeople, with the visual culture of entertainment gaining an increasingly strong presence.

Packaging Pleasure

As described in chapter 2, children's software often relies on educational claims in being marketed to achievement-oriented parents. Yet as the market for children's software developed, a larger proportion of titles looked to entertainment as the primary goal, with curriculum-based products gradually ghettoized as "educational" and by implication not as fun as other software available on the market. Michaud commented astutely on the shift away from educational content. "A lot of how well your product is received in the market is dependent on how well you position it as educationally entertaining." She reflected: "Probably the biggest change in the industry is the acceptance of parents and teachers alike that games don't have to have that much learning content anymore." I pushed her on this point: "Why is that?" Michaud thought a moment and then answered with a description of the state of the industry at that particular time (1999):

Because the kids have to actually like playing the game; otherwise, the parents feel as though they've wasted their money. Three years ago, when there wasn't much competition in the educational software market, content was king. Products were expected to have a lot of learning content. And if it was fun, that was a bonus. The fun was secondary to the child doing his three Rs. Now, with so many different software games to choose from, kids really have to like the software they play. Kids today are technically savvy. They're learning to use computers and game consoles at an early age so their standards are much higher for what they're entertained by. I think parents are savvier, too. So entertainment is more important now. I'm amazed at products that were at the top of the charts three years ago, compared to what's at the top of the charts now. Three years ago it was *Oregon Trail*, it was *Learning Company Math*—that kind of thing. Now it's Mattel's *Barbie Fashion Designer*.

Her feeling was that "the trend is really towards highly entertaining software, and educational content is becoming increasingly secondary." She saw Humongous Entertainment, creators of *Pajama Sam* and *Freddi Fish*, as a case in point.

Humongous, when it first came out, was kind of looked down upon by teachers because it didn't have any content in it. It was just a fun game for kids. Now, the various Humongous product series are accepted as legitimate learning products both in school and in the home. Critical thinking, if that's the only content you have in your game, it's enough. What is critical thinking? It's become this obscure, broad-based buzzword that everybody puts on their package that is practically synonymous with the word *educational*.

The packaging and marketing of many of the popular titles on the market today supports Michaud's comments. Since the late 1990s, children's software has been characterized by visually polished multimedia titles that can compete with entertainment media in terms of production value. The market for children's software is polarized between curricular products based on a pastiche of school-coded content and "wholesome entertainment" titles that are marketed as an alternative to video gaming, providing fun and excitement without the violent content and play mechanics of action games. At either of these poles, a certain level of graphical appeal is a basic requirement, but the two kinds of products rely on different selling points, education or entertainment.

In *Great Software for Kids and Parents,* a book in the Dummies Guide to Family Computing series, Cathy Miranker and Alison Elliot describe what they didn't want from their family computer. "We didn't want it to turn into a high-priced video-game player. We didn't want it to be an electronic baby-sitter. We didn't want it to be a desktop TV"; instead, they were looking for software that

• Creates inventive, hands-on opportunities for fun learning;
• Offers kids something special, something that takes advantage of the computer's unique capabilities;
• Encourages kids (and parents) to make connections between their computer-inspired discoveries and real life;
• Fits in with our kids' lives without eclipsing the books and toys, games and adventures, and traditional pastimes that we value. (1996, 2)

Miranker and Elliot's orientation encapsulates the approach many tech-savvy parents took in the late 1990s during the heyday of edutainment. The software needs to be fun and engaging, but different from the passive and lowbrow media of television and video games, and, it is hoped, educationally enriching from a curricular standpoint, although this last quality is not necessary. Rather than take a narrow view that software is valuable only if pushing academic content, the two authors look more holistically at how the activities fit into the overall ecology of home life and play as well as into academics. They review both curricular products and entertainment products such as *Barbie Fashion Designer* and *SimCity.* The hefty price tag of early entertainment multimedia products and the stance toward childhood enrichment in this genre fell in line with the middle-class concerted cultivation that Lareau (2003) describes and that characterizes the academic genre. What makes the family entertainment genre different, however, is the openness to more accessible and lowbrow entertainment idioms and a more permissive approach to childhood pleasures.

The ads for the more entertainment-oriented titles portray children as ecstatic and pleasure seeking rather than reflective and brainy, and depict childhood as imaginative, pure, and joyous. The ethos is parent-friendly but child centered, a formula established by children's media companies ever since the Mickey Mouse Club aired on television. Rather than playing on achievement anxiety, ads for these kinds of titles play on parents' desires to indulge their children's pleasures and on the growing pressure on parents to be in tune with their children and keep them happy and entertained. A child's happiness has become as much a marker of good parenting as achievement and effective discipline.

In contrast to the generic and faceless child facing the refrigerator in the *Jump Start* ad described in chapter 2, a Humongous Entertainment ad puts the child's pleasure front and center. "This is the review we value most," declares the ad copy above a large photograph of a beaming child (figure 3.5). Humongous's adventure game *Putt-Putt Joins the Circus* does not make specific curricular claims other than promising an engaging and prosocial orientation. The ad lists "problem solving, kindness, teamwork, friendship" as the game's educational content items. It mobilizes discourses from the established genre of film reviews by describing how "critics rave" over the software. The bottom of the ad lists quotes from various reviews in software magazines. The last quote, from *PC Magazine* is particularly telling: "Nobody understands kids like Humongous Entertainment." The company is thus positioned as a channel to your children and their pleasures, the authentic voice of childhood.

The box for *Pajama Sam*, one of Humongous's most popular titles (figure 3.6), features the adorably caped hero, Sam, and describes the software as "an interactive animated adventure." The back of the box does list educational content in a small inset that is visually decentered from the portions describing the excitement and adventure that the title promises. The list—including "critical thinking, problem-solving skills, memory skills, mental mapping and spatial relations skills"—does not make any curricular claims and stresses the "creative and flexible" nature of the software and "the power of a child's imagination." "Feature-film quality animation" and "original music" are central selling points for the title. It thus can compete

Figure 3.5

Advertisement for *Putt-Putt*. Putt-Putt® Joins the Circus™ and Pajama Sam® artwork courtesy of Infogrames Interactive, Inc. © 2002 Humongous Entertainment, a division of Infogrames, Inc. All rights reserved. Used with permission.

with television and videos for your child's attention, but still has some educational value. Children's "natural" imaginations and creativity achieve full expression through the mediation of sophisticated media technologies and visual culture.

"Prepare to get blown away!!" screams the copy above a wide-eyed boy dangling off the edge of his PC as if he is hanging from a cliff. "The action in Disney's CD-ROM games is so awesome, your kids are gonna freak (and that's a good thing). So hold on tight and check out the action this holiday!" As far as wild fantasy goes, these products are relatively tame, based on familiar Disney formulas of fast-paced adventure and gore-free violence. Yet the pitch highlights the software's action and "freaky" aspects as its primary appeal. Although still addressing the parent, the ad copy makes use of children's language, hailing the hip parent who is in touch with children's culture and desires. In contrast to the conservative apparel worn by the *JumpStart* children, this boy is dressed in baggy skate-punk shorts and trendy sneakers, and has spiked hair with blond highlights. The children in this ad and the Humongous ad are white and presumably belong to middle-class PC-owning families, but they are not marked as academic achievers. In this ad, the vernaculars of children's peer and popular cultures are mobilized to enlist the progressive parent and to position Disney as the authentic voice for children.

The LEGO Company similarly features children's pleasures in an ad entitled "Imaginations Powered Daily." "Let your star shine" declares their girls' product *LEGO Friends* (figure 3.7a). "His own LEGOLAND theme park!" suggests the *LEGOLAND* copy below a beaming boy holding a blueprint (figure 3.7b). "Let their imagination run totally wild as they are challenged to build the LEGOLAND of their dreams." Ads for another LEGO title also promise excitement and fun. "His license to thrill!" proclaims *LEGO Racers*, featuring a too-cool boy in shades and car-racer garb (figure 3.7c). "Rock his world!" shouts the copy on *LEGO Rock Raiders* (figure 3.7d). This ad campaign promises to cut children's imaginations free of their everyday constraints and responsibilities, letting them run wild in thrilling, action-packed, online adventure. The LEGO Company tames adult and teen-oriented video game culture into an entertainment vernacular that still preserves a protected space of childhood innocence. Unlike mainstream video games, the titles are still marketed to parents, and the children are depicted in adorably oversize costumes. The ads assure parents

a

Figure 3.6

(a) Front cover and (b) back of the box for *Pajama Sam 2*. Putt-Putt® Joins the Circus™ and Pajama Sam® artwork courtesy of Infogrames Interactive, Inc. © 2002 Humongous Entertainment, a division of Infogrames, Inc. All rights reserved. Used with permission.

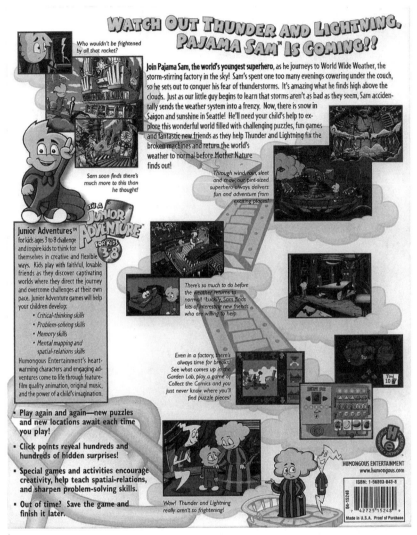

b

Figure 3.6

(Continued)

a   b

Figure 3.7

Advertisements for LEGO Group Products: (a) *LEGO Friends,* (b) *LEGOLAND,* (c) *LEGO Racers,* and (d) *LEGO Rock Raiders.* LEGO, the LEGO logo, and LEGOLAND are trademarks of the LEGO group. © 1999 The LEGO Group. Images used here with permission. The LEGO Group does not sponsor or endorse this book.

c   d

Figure 3.7
(Continued)

that despite these adult fantasies and the aspirations to youth culture and action media, children are, after all, just children.

These more entertainment-oriented titles use the same visual elements as the academic titles described in the previous chapter. Both academic and entertainment titles share the same stylistic genre, and many titles are not clearly categorized as one or the other. They occupy the same shelves at retailers and are oriented to a similar demographic of middle- and upper-middle-class families, but the entertainment titles are keyed somewhat toward the more progressive and permissive parent. What distinguishes entertainment as a genre is the orientation toward more indulgent and repetitious play in contrast to competitive and linear progress. Academic titles, in particular those that make curricular claims, are generally linear and make much of achieving certain levels and scores. By contrast, entertainment software is exploratory, often repetitive, and generally open-ended. What gets packaged and marketed in this software is not achievement, but fun, exploration, and imagination. These titles are also distinguished from the action-entertainment titles marketed primarily toward teens and adults. In contrast to the darker hues and often frightening characters adorning the action software boxes, brighter colors and smiling, wide-eyed characters like Pajama Sam clearly code entertainment software for younger children as a separate market.

Entertainment-oriented children's software is a comparatively broad genre. Although one can easily recognize the academic nature of a series such as *JumpStart* or *Math Blaster,* entertainment titles can include software such as authoring tools, interactive storybooks, or simulations, and the box design and marketing pitches are less uniform. Most typically, titles that are specifically in the children's software category of entertainment feature an open-ended and exploratory online environment for players to explore. They may feature gamelike idioms, but are often more open-ended and toylike, and they tend toward sandbox-style games rather than linear, goal-oriented ones. One software title that was popular in the 5thD during my period of fieldwork exemplifies many features of this genre.

Software Case: The Magic School Bus Explores the Human Body

When I was completing my fieldwork at the 5thD at the end of the 1990s, CD-ROM games and a more entertainment-oriented genre of products were making their appearance at the clubs. Mainstream licenses such as LEGO, Barbie, and Disney were yet to arrive at the children's software scene, so I was not able to see such titles in my play settings. We were just beginning to see the emergence of licensing arrangements and tie-ins with television and other media, as well as more and more titles with CD-ROM-quality production value. Broderbund's Living Books series, licensing popular children's books such as those by Mercer Mayer *(Just Grandma and Me)*, represented one such tie-in. Another, which I examine in more detail here, is the Magic School Bus series of CD-ROMs.

The Magic School Bus series of games is an adaptation of the Living Books interactive storybook format and is based on a popular children's book series and television series. The games were produced by an alliance between Scholastic, the publisher of the book series, and Microsoft—representing Microsoft's first major foray into educational publishing and Scholastic's first foray into software. The series received much attention in the press and favorable reviews (Ruocco and Dyson 1996). *The Magic School Bus Explores the Human Body* (*MSBEHB*) was the second in a series of eight titles to date. With their school-like content, these titles incorporate many elements of edutainment, but they are evidence of a shift toward the vernaculars of TV-centered popular culture. The series represents a transitional moment in the shift from academic to entertainment as the dominant genre in children's software. Academic content is still a focus in the series, unlike *Pajama Sam,* which has little overtly school-like references.

The packaging of the software exhibits the emerging orientation toward the wackier, visual, child-centered media culture of children's television. The cover (figure 3.8a) declares a "fun-filled, fact-packed science adventure!" The back of the box (figure 3.8b) marks and separates out the appeals to children and to parents, with the appeal to kids featured more prominently. "Hey kids!" calls out one of the characters. "There's fun ahead. In the front, back or outside the bus! Just click and you can see what is going on inside you." "Hey parents!" another character calls out. "Here's why exploring the human body in Scholastic's The Magic School Bus® is absorbing for kids and can be a great part of their diet of activities!" The box then lists the twelve body parts introduced in the software and sings the praises of multimedia. "Science facts come to life with narrative, sound, video and animation."

a

Figure 3.8
(a) The front and (b) back of the box for *The Magic School Bus Explores the Human Body*. Reproduced with permission from Microsoft Corporation.

MSBEHB is a multimedia space of exploration organized by a fantasy scenario of traveling through the human body. The narrative logic is one of free-wheeling and noncompetitive exploration. The teacher, Ms. Frizzle, and her magic bus invert the power dynamics of the traditional classroom. The adventures depict kids escaping the disciplines of formal education, adventuring beyond the classroom walls to embark on fantastic adventures with a slightly crazy and out-of-control teacher and a shape-shifting school bus. The kids are often the more level-headed and calmer characters, strug-

b

Figure 3.8
(Continued)

gling to keep up with their charismatic leader. Like the entertainment industry content creators, Ms. Frizzle stands in for the liberated adult who is in touch with her uninhibited, playful, inner child. Rather than being presented with the linear and progressive logic of the classroom curriculum, subjects such as the human body, space, and geology are explored through a chaotic and dizzying set of encounters where the characters in the story careen from one scene to another.

The player enters the game through the classroom of Ms. Frizzle and can click and explore various animated objects in the classroom. The software draws from the interactive idioms developed in earlier titles such as *Silly Noisy House* and in the Living Books series. For example, clicking on a fishbowl will make the fish jump up, and clicking on a model volcano will make it erupt. A skeleton in one corner turns into a puzzle that the player has to reassemble. The main adventure is triggered by clicking on the toy school bus in the classroom, which launches a movie in which the player enters one kid's mouth along with a handful of cheese puffs. From there, the player can visit twelve other parts of the body, such as the liver, lungs, esophagus, stomach, and intestines. In each area of the body, the player is accompanied by the cartoon characters from the series—Ms. Frizzle and her students—and can hear them talk or report on the particular body part. Other activities include conducting a virtual science experiment, using a drawing application, clicking on various objects to get animations somewhat related to the body part, and playing a video game loosely thematized around the given part of the body.

The dominant way of interacting with the game is "click and explore" and is based on a relatively intuitive interface and navigation mechanic. The player clicks on road maps to move to a different part of the body and on an object to activate it. Most objects in the game have no functionality other than triggering an animation or a scene change. One exception is the set of minigames triggered by clicking on the hand-held game machine that one of the kids is playing. These games are generally very technically simple, borrowing from existing models of early video games. For example, for the nose, the player has to put together a puzzle, and in the lungs the player can access a game of pinball. The title also has a toolbox in certain scenes that allows the player to view scenes from different perspectives, such as with X-ray vision or by means of a flashlight. Help is embedded in the game through "Liz," a small green lizard that is always present at the bottom of the screen. Liz often gives advice on navigation and other aspects of game functionality, and offers minilessons on the body.

The organization of tasks is integrated with the fantasy scenario and content domain. For example, "street signs" for navigation follow the layout of the human body by moving from mouth to esophagus to stomach to small intestine, and so on (figure 3.9). The tasks are organized around a coherent though fantastic story of being in a tiny school bus that is traveling through a body. The only elements not clearly integrated into this fantasy scenario are the video game portions, which are often inciden-

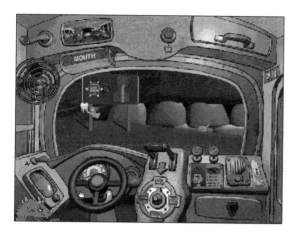

Figure 3.9
Screen shot of *The Magic School Bus Explores the Human Body.* Reproduced with permission from Microsoft Corporation.

tal to the body theme. Although the video games break the overall narrative logic because they are peripheral, the design still retains coherence between structure and fantasy scenario. In contrast to *Dr. Brain,* where the puzzles and games were the primary focus for interaction, the "game breaks" in *MSBEHB* are connected to the fantasy scenario of exploration.

The software provides no particular sequence for clicking on different parts of the classroom or other scenes, and once in the body, the "nervous system navigator" allows the player random access to any other body part. There are also no right or wrong answers, but simply information and action triggered by mouse clicks. Other than the video games, there is no way to measure achievement according to a goal-like metric, except for when the player exits the game and gets a "passport" stamped according to the areas of the body visited, but not according to the activities engaged with in any given area. The passport is more of a tally for the player to know how much of the software he or she has engaged with, rather than a report card of achievement. There are no levels or scores in navigating the body, just different places to travel to and explore. *MSBEHB* is clearly an adventure, not a test.

Although the player has a great deal of flexibility in determining pathways through the scenario, she or he has little ability to input content or to interact beyond a single mouse click. The structure of the software parallels the entertainment orientation of noninteractive media by providing content that is almost entirely preprogrammed and waiting to be activated by the user. One mouse click is often followed by a long animation sequence, character dialog, or a transitional sequence. The game relies on a "spectator" position more than on a personalized and interactive "student" position. The most technically sophisticated aspects are not the mechanics of game play or user authoring, but rather graphic and sounds. Much of the animation and graphics have high production value and clearly are the areas in which the majority of design effort was invested. The sounds have been designed to be appealing to kids, with many sound effects and "gross" sounds in almost every scene. In contrast to simulation games that stress user authoring and academic games that assess a child's knowledge, *MSBEHB* is more like an engaging and animated book that can be opened at any point but doesn't demand much from the user. This approach contrasts to the marginalization of the fantasy and exploratory elements in *Dr. Brain* and grade-based products, where one scene is a precondition for progressing to another. Other than in its video game breaks, *MSBEHB* is noncompetitive and nonevaluative, driven by a narrative and spectacular logic rather than by a gamelike or achievement-oriented one.

MSBEHB in the 5thD

MSBEHB was popular in the 5thD in the period after its introduction in 1995. This popularity may in part have been due to the fact that it was introduced to the club with fanfare as a gift from the Wizard, but also to its flashy graphics and sounds, which seemed to catch kids' attention as they wandered about club. Every tape we made shows a substantial revolving audience of different kids and undergraduates checking in on the game play, especially in sequences where the game produced particularly gross or funny sounds. *MSBEHB* found a place in the 5thD maze very quickly, and many kids played it based on the location in the maze sequence.

Although most undergraduates did not make any particular comments regarding their impressions of the game, and some expressed confusion about the point of it, many were enthusiastic in describing it. Their field notes say that MSBEHB is "pretty cool," "amazing," "really neat," and "fascinating." Some point to the "great" or "cool" graphics. One undergraduate was even more enthusiastic: "I think this is the best computer game I have ever seen! It is visually stunning and completely enthralling. The sound effects are great, a bit on the vulgar side, but I suppose that keeps kids' attention pretty well! It is a completely educational game—there is no way that you cannot learn something from it!"

The kids' descriptions of the game seemed to split along age lines. Younger kids repeatedly described it as "cool" and "fun." Older kids, however, viewed it as a "kiddie game." As a twelve-year-old kid was playing it, a group of older boys walked by and called it "The Magic Retard Bus." Similarly, the twelve-year-old who was playing it made sarcastic remarks about its content, calling one of the characters "a fag" and mimicking some of the talk of the characters in an annoying and childlike voice. In a field note, an undergraduate records how a twelve-year-old girl "thought that the things they told/taught you were kind of elementary, not at her level." The title was in general not described as "easy" or "hard," except for its video game portions.

A number of features of play in *MSBEHB* make it distinct from *Dr. Brain* and *SimCity 2000,* which appeared after it. Unlike with *Dr. Brain,* for this game there was no focused period of videotaping; instead, it appears regularly in our video record throughout the central year of observation. A game such as *MSBEHB* does not invite a focused, goal-oriented engagement, and kids tended to engage with it in sporadic and more lightweight ways. Because it was embedded in the maze as part of 5thD's activity system, many children played it but were not necessarily invested in it the same way that children playing games such as *Dr. Brain* or *SimCity* were. So we have many observations of play with the title, but we do not have sequences of particular children playing across multiple days—a consequence of the game's location in the 5thD ecology as much as of its design. The remainder of this chapter ranges beyond examples of *MSBEHB* play to illustrate processes of play with the entertainment genre of software in general, and there is no sustained case study of a particular child's subject formation in relation to the software. In the previous chapter, *Dr. Brain* was the focal case for examining the content of as well as play in the edutainment genre. Here I describe play with the narrative features particular to *MSBEHB* and then turn to aspects of play with the entertainment genre that were evident in the 5thD across a variety of software titles: engagement with special effects, discourses of "fun," and narratives of destruction and action entertainment.

Narrative and Exploration in *MSBEHB*

MSBEHB foregrounds narrative logic and visual appeal over technical sophistication or user authoring. At the 5thD, these narrative and visual features—the educational information, Magic School Bus backstory, movies, and sound—became the focus for both kid and undergraduate engagement. Elements such as the video games, the drawing program, and the science experiments afford some peripheral and more player-driven forms of engagement. The player experience is designed for engagement primarily as a reader-viewer, albeit one with more control over the story line than is the case with a traditional book or movie format.

During moments of play, there were ongoing negotiations between kids and undergraduates about what narrative trajectory to take through the software and how to engage with the educational content. One persistent tension occurred between undergraduates, who tried to orient kids to the educational content and systematic exploration of all areas, and kids, who were oriented toward the visual effects and a more chaotic trajectory. In almost every field note about the title, undergraduates describe their efforts to get kids to engage with the content domain or to get kids to click on various aspects of a scene before zooming off to the next body part. One undergraduate, after working with a kid for the second time, wrote, "I asked if maybe this time through he thought we could explore each area in more depth. This added to the game, because instead of just visiting different places in the body, we explored them as fully as possible."

In one instance of play we recorded, a boy, John, is working with two undergraduates, Peggy and Elaine. John is controlling the mouse and has been exploring various parts of the human body. Throughout the interaction, John clicks around the scene as the undergraduates work to engage with him and the educational content of the game. They suggest exploring an area further before zooming off to the next body part, or they ask him if he knows what an intestine is. In this excerpt from the video, they all are observing an animation of the bus traveling through an artery, and both undergraduates work to create a dialog on blood and blood cells, specifically invoking school learning in the process.

P = Peggy
E = Elaine
J = John
C = Computer

1   P: What do you think these things are? (Referring to disclike flying objects.) Those red things.

2   J: Um, they're, um, blood.

3   P: Yeah, blood cells.

4   E: OK now, if you click on that, I think it might tell you something about the kidney.

5   C: Many people can live with one kidney. If one of the kidneys is damaged, it can be removed, and the remaining kidney will do the work of both.

6   P: Try exploring. Click around all over the place.

7   J: (Clicks on another place, and an animation of traveling through a vein starts.)

8   P: So where do you think he's traveling in if those are little blood cells? How is he getting from place to place? What are these tunnels? Do you know what those are?

9   J: Those are your water input, right? (Gestures along throat.)

10   P: OK, what do you have right here? (Points to J's hand.)

11   J: Veins.

12   P: Yeah, veins. And you have arteries. That's what carries the blood. Have you learned about the human body yet? In school?

13   J: No. (Animation ends, and bus cockpit appears. J starts clicking on objects outside of windshield, resulting in humorous animations and sounds.)

14   P: See, they're blue blood cells.

15   J: What is that?

16   E: I think that means it needs to get oxygen.

17   J: White blood.

18   P: White blood cells?

19   E: You want to see what's kind of neat? Try clicking on the mirror.

Upon seeing the animation, which includes representations of flying red blood cells, Peggy asks John if he knows what they are (line 1), and he responds with a partial answer (line 2). The interaction is akin to a student-teacher interaction. Elaine suggests clicking on an icon (line 4), which brings up an animation and a narration describing the functions of the kidneys (line 5). After John clicks on another icon, and they see the animation of the arteries again, Peggy again resumes her questioning (line 8) and, after one incorrect response (line 9), gets the correct answer from John, then asks him if he has learned about the human body in school yet (line 12). Their dialog is interrupted by a new animation appearing on the screen, and John begins to click on different objects, triggering more animations. For example, a white blood cell animates into an ambulance when clicked on. Elaine suggests clicking on a different part of the bus (line 19), which takes them to a different screen. In a field note, Elaine glowingly describes this afternoon as a successful day of interaction with a child. She writes that John is "diligent" and "polite" and that she is impressed by his ability to read and follow instructions. "He had no problems in talking to both Peggy and I, and he accepted our help willingly. He never specifically asked for our help but he never refused it either." Peggy's field note is similarly positive about the engagement.

More typical was a degree of frustration about the difficulties of getting kids to respond to adult intervention. For example, one boy declared flatly, "I don't want to learn," when an undergraduate tried to get him to open and read the informational drawer. Another undergraduate describes in her notes how the site director intervened in their play, guiding both the undergraduate helpers and the kids to engage with questions about the human body: "He came over and tried to prompt discussion by asking questions about what things do in the body, or what the drawing was representing. This led to more questions. However, for the most part, the kids did not appear interested in answering our questions. They just wanted to play the game."

Although undergraduates were impressed in a few instances with kids' interest and knowledge about the human body, they wrote about many more instances of frustration with the kids' unwillingness to engage beyond the spectacular features of the software. One undergraduate describes her difficulties in working with one boy: "[We] explored many places in the body together, but it was difficult to try to get him to focus on what we were doing, or the purpose of it. He was much more interested in having the pieces of semi-digested food in the stomach make their burping/farting noises than what purpose the stomach actually served. Each place we went to, I asked him what purpose it served, but he just wanted to plunge ahead."

Another undergraduate tells how she got "really frustrated" with one girl who would not respond to her offers to help her play and to talk about the educational content. "I kept trying with each different body part and she started to answer me. But when she did, she answered that she didn't care, and that became very obvious." The transcript of this session is indicative of a tension between the undergraduate's agenda and the kid's agenda.

UG = Undergraduate
M = Mary

1   UG: What happens if you hit the brown thing?
2   M: (Hits the sign to the small intestine.) To the small intestine.
3   UG: What does the small intestine do?
4   M: I don't know.
5   UG: Then we can find out, right?
6   M: Eww!
7   UG: What's that?
8   M: (Unintelligible) That's the lower stomach here.
9   UG: That's why yours does too. All right here.
10   (They arrive at the small intestine.)
11   UG: It goes all through here.
12   (M clicks on some of the floating things.)
13   UG: What happens if you point at it? It disappeared. Huh, look, it is fat. But you don't know what the small intestine does yet.
14   M: (Clicks the sign to the large intestine.) Who cares?
15   UG: Don't you want to learn what your body does?
16   M: Oh, look, there's some Cheetos right there.
17   UG: Mm-hmm.
18   M: Eww. What is up?
19   UG: You're on your way to the large intestine. Do you know what the large intestine does?
20   M: No.
21   UG: Do you want to find out? You don't have to if you don't want to.

The undergraduate struggles to get Mary to take an interest in what a small intestine is, repeatedly trying to engage her: "What does the small intestine do?" (line 3), "But you don't know what the small intestine does yet" (line 13), and "Don't you want to know what your body does?" (line 15). Mary responds bluntly, "Who cares?" to the second statement and points the undergraduate to the Cheetos floating in the intestines. She is more interested in the gross visuals, repeatedly calling attention to them with an appreciative "Eeew!" Undergraduates experienced frustration and a feeling of being out of control when kids were unresponsive to them in this way and insisted on pursuing their own agenda of seemingly random engagement with the game's surface appearances.

Although the kids exhibited different degrees of tolerance and responsiveness toward the undergraduates' attempts, they always demonstrated an orientation toward the game's graphics and special effects rather than toward its content. In the 5thD context, where the undergraduates were instructed to keep engagement "fun" and child centered, the kids were generally successful in acting on these preferences. The kids could spend excruciatingly long minutes playing with gross farting and squirting noises, despite (or perhaps because of) the undergraduates' discomfort. Younger kids who were familiar with the *Magic School Bus* TV series also showed interest in the characters and the overall Magic School Bus story.

Storybook games such as *MSBEHB* rely heavily on the uniquely spectacular features of interactive multimedia to draw children into engagement with educational content. The basic mode of play is to click on objects to see what visual and auditory effect will result. Kids find these small animations highly amusing, particularly if they are accompanied by a gross noise or visual. One undergraduate describes how two girls working together wanted to click on every animated object in the classroom. One of the girls said she liked the game because of "the music and the weird sounds it makes and how you can go into the human body." The sounds often drew an appreciative audience and "advertised" the game to other kids at the club. One undergraduate notes how the girl she was working with "at first only wanted to learn how to produce those 'cool noises.'" Although other kids were not so systematic in clicking on every object, they did enjoy the animations and showing off particularly "cool" ones they had discovered. In the tapes where kids are together at the game, kid observers are constantly leaning in and pointing at things to click on. In the following sequence, Peter is in control of the mouse, and Chris and Brad encourage him to click on different things (lines 1–3, 6, 8):

C = Chris
P = Peter
B = Brad

1   C: Get that—get that thing. Get that one candy bar. Get the candy bar. Get the candy. Get the candy bar. Get the small chocolate thing. (Leans in and points). Awww. You should've got it. It's funny.

2   B: Get that one. (Leans in and points.) This one's funny.

3   C: Oh, get that one thing. Right here.

4   P: (Clicks on blob that turns into a hamburger with an audible yawn.)

5   B: Oh yeah. Cool.

6	C:	Get that chocolate thing. Hit it.
7	P:	(Hits blob that turns into a chocolate bar.)
8	B:	What's that green thing?

The kids rarely tired of this mode of clicking on animated objects and revisited areas to show particularly "cool" interactions to other kids and their undergraduate helpers. This interaction was more a version of show-and-tell than the displays of mastery evident with *Dr. Brain.*

Mastery of the software was defined by knowledge of the interface and navigation rather than by knowledge of the content domain. When children leaned in to offer suggestions and help, their comments were exclusively about such things as what to click on to get a cool special effect or how to get to a specific place in the scenario. Although undergraduates viewed knowledge of the human body as an important aspect of engaging with the title, that knowledge was not portrayed as central to "knowing how to play." Even undergraduate descriptions of their own lack of mastery of the software similarly revolve around not knowing what to click on rather than knowledge of the educational content. One undergraduate describes her sense of being lost and "not having authority" because she didn't know where to click to get to different scenes of the game.

In summary, play with *MSBEHB* at the 5thD was dominated by the open-ended narrative of exploration and discovery rather than by a focused trajectory, achievement, or content mastery. Adults made frequent attempts to orient kids toward a more systematic and progress-oriented form of engagement, but the kids tended to resist, particularly when several of them were playing together and sharing their discoveries and fascinations. Kids' oriented themselves instead to visual and auditory discovery and a more chaotic trajectory through the narrative space. For them, key factors in the software's appeal were the spectacular multimedia dimensions. I turn now to a discussion of the logic of special effects in this and other graphically sophisticated multimedia titles.

Spectacle and Special Effects

Here I present cases from my ethnographic record of play with *MSBEHB,* *SimCity 2000,* and *DinoPark Tycoon,* focusing on players' engagement with the visual, auditory, and interactional special effects. These games embed multiple goal structures and invite different forms of engagement, but are similar in that they all are CD-ROM games with high production value and polished graphical and auditory effects.

Visual Effects

The tapes of kids' game play with graphically advanced games at the 5thD are continuously punctuated by their notice of on-screen eye candy, an occasional "cool" or "oooh" that testifies to their appreciation of visual aesthetics of one kind or another. One undergraduate describes a boy playing *SimCity 2000* for the first time. "Every time he placed a building on the screen, he exclaimed 'Cool!' because the graphics were very complex and vivid." With a title such as *MSBEHB*, attention-grabbing graphics are central to the game's appeal because the game relies on an exploratory mode of interaction rather than on a strong narrative story line or a competitive goal orientation. The animations that form the transitions between the different parts of the body often drew appreciative "EEEWs" from both undergraduate and kid viewers, as they watched the tiny bus drop into a puddle of stomach goo or fly down a sticky esophagus. "This is the fun part. This is fun. Watch," insisted one kid as he initiated the opening animation. An undergraduate notes how a girl was "really excited" about showing her one small animation in *MSBEHB*.

The screen for designing a face to go on the driver's license in *MSBEHB* invited many minutes scrolling through the different options for facial features and discussion of what is a cool or uncool feature. When multiple kids were engaging with the software together, there invariably was extended discussion about and exploration of different visual features in the driver's license. For example, the video shows two boys working together on a game and arguing about each facial feature, such as the eyes and eye wear (lines 4–7) or skin color (lines 8, 9).

C = Chris
P = Peter
UG = Undergraduate

1   C: We're going to do the same things as last time, like you want. (Scrolls through features.)
2   P: Yeah, that.
3   C: That one you did, you wanted. That was a good one.
4   UG: Change the eyes.

5   C: I will. I'm going to do the one you wanted.

6   P: Yeah, I like that one. No, put the really thick sunglasses. Go back.

7   C: No, I like those. I like those sunglasses.

8   P: Then I get to pick. Change skin color. Do you see a difference? Maybe it just gets it darker.

9   C: There. Like that? (Completes face.)

Features such as the driver's license, animations of things in the different scenes, or different controls in the cockpit provide an ongoing stream of visual effects that are often irrelevant to the educational content, but provide eye-catching distractions that keep kids engaged. *MSBEHB* incorporates visually spectacular features that are ends in themselves for players, regardless of their relevance to the central play action. *SimCity 2000* was another graphically advanced game at the site during the period of our observations. Although it did not offer the same appeal to the grotesque that *MSBEHB* did, it, too, invited pleasure in the visually spectacular.

J = Jimmy
H = Holly

1   J: I want to do a highway (selects highway tool). How do I do a highway? OK. (Moves cursor around.) I'll do a highway right here.

2   H: Right there? I think you should have it . . . hmm . . . trying to think where a good place for it . . .

3   J: Right here? Here? (Moves cursor around.) Here? (Looks at Holly.)

4   H: Sure. What is that place there, residential?

5   J: (Budget window comes up, and Jimmy dismisses it.) Yeah. I'm going to bulldoze a sky-rise here. (Selects bulldozer tool and destroys building.) OK. (Looks at H.) Ummm! OK, wait, OK. Should I do it right here?

6   H: Sure, that might work. . . that way. Mmmm. You can have it. . .

7   J: (Builds highway around city.) I wonder if you can make them turn. (Builds highway curving around one corner.) Yeah, okay.

8   H: You remember, you want the highway to be . . . faster than just getting on regular streets. So maybe you should have it go through some parts.

9   J: (Dismisses budget. Points to screen.) That's cool! (Inaudible) I can make it above?

10   H: Above some places, I think. I don't know if they'd let you, maybe not.

11   J: (Moves cursor over large skyscraper.) That's so cool!

12   H: Is that a high-rise?

13   J: Yeah. I love them.

14   H: Is it constantly changing, the city? Is it like . . .

15   J: (Builds complicated highway intersection. Looks at Holly.)

16   H: (Laughs.)

17   J: So cool! (Builds more highway grids in area, creating a complex overlap of four intersections.)

18   H: My gosh, you're going to have those poor drivers going around in circles.

19   J: I'm going to erase that all. I don't like that, OK. (Bulldozes highway system and blows up a building in process.) Ohhh . . .

20   H: Did you just blow up something else?

21   J: Yeah. (Laughs.)

22   H: (Laughs.)

23   J: I'm going to start a new city. I don't understand this one. I'm going to start with highways. (Quits without saving city.)

This sequence during a child's (Jimmy) play with an undergraduate (Holly) is punctuated by moments of engagement with the interface as visual special effect. At a certain point in the game, as his city grows, Jimmy attempts to build highways. "I want to do a highway," he declares, selecting the highway tool. "How do I do a highway?" (line 1). Moving his cursor around, he discusses with Holly where he might put the highway, settling on an area near a commercial district (line 3). He bulldozes to make way for the highway and then builds it around one edge of the city, discovering at a certain point that he can make it curve around the corner if he clicks on blocks perpendicular to one another (lines 5–7). As he builds his highway in the foreground, he notices that it is elevated above the level of the regular roadways. "That's cool!" he exclaims. "I can make it above?" (line 9). Holly speculates on whether they can build the highway through the city, and then Jimmy points with his cursor to a tall, blue-and-white skyscraper: "That's so cool!" (lines 10, 11). Holly asks, "Is that a high-rise?" (line 12). "Yeah," Jimmy answers. "I love them," he declares emphatically (line 13). Jimmy goes on to continue his highway and then discovers that if he makes overlapping segments, they result in a cloverleaf. He looks over at Holly with delight when this happens, and she laughs. "So cool!" he exclaims, building further overlapping segments that result in a twisted quadruple cloverleaf (lines 15–17). "My gosh," says Holly, "you're going to have those poor drivers going around in circles" (line 18). Jimmy then bulldozes the whole cloverleaf pattern, blowing up a large building in the

Figure 3.10
Jimmy's highway to nowhere in *SimCity 2000*. Screen shot reproduced with permission from Electronic Arts Inc. © 1993–1994 Electronic Arts Inc. SimCity 2000 and SimCity are trademarks or registered trademarks of Electronic Arts Inc. in the United States and/or other countries. All rights reserved.

process (figure 3.10), and then declares that he is going to start a new city (line 23). He closes his city without saving it.

Although this sequence begins with certain accountabilities to building a transportation system, by the end Jimmy has wasted thousands of dollars on a highway to nowhere, blown up a building, and trashed his city. Holly draws him back into the accountabilities of building a well-functioning city by pointing out that the highway cloverleaf might look cool, but is not going to work very well. Her intervention is subtle, but it has the effect of calling him away from spectacular engagement to the game's more functional accountabilities. Jimmy responds to her suggestion by trying to fix the highway, but eventually decides to start over because he has wasted too much money on playing with the highway as special effect. He apparently has few attachments to the city that he has worked on for more than thirty minutes and, in fact, replicates a pattern of building up cities to a point of difficulty and then getting rid of them, not bothering to save or follow up on his work.

A game like *SimCity 2000* provides visual rewards that are linked to but not isomorphic with the more functional rewards of building a large, financially stable, and well-populated city. As the city grows, the player is given more and more visually stunning buildings, such as stadiums, marinas, and space-age buildings. "Oh wow. Look at that!" an adult helper called out in one case after they finished building an airport. "I got a helicopter! Girl, I got a helicopter," said the girl he was working with, calling out to her friend. Kids who were invested in the game developed a certain connoisseurship about the placement of visual features. This same pair took pains to build a series of marinas and waterfalls, positioning them just right, so that the tiny sailboats did not fall down the waterfalls, and destroying a prison that they felt was too close to this area. In another instance of play, one kid built a large pyramid-shaped structure covered in water. "I built this by hand," he told his undergraduate helper with pride. The pleasure here is in a certain personalized aesthetic, in the spectacle of the interface, not in the instrumental goals of the game or in inhabiting an engaging virtual reality. It is in the visual special effect rendered in interactive media.

Interactional and Auditory Special Effects
Unlike media such as film and television that are the targets of Guy Debord's critique of passive consumption, interactive media are predicated on the consumer's active engagement. Rather than negating the medium's spectacular qualities, this interactivity actually serves to create a new genre of special effect in which the player is able to control and manipulate the production of the effect. This is clearly evident in "twitch" games that demand close sensory-motor coupling with visual and auditory effects. In the games that form the basis for this study, these interactional effects are still present, though not as central as in action gaming. Although visual effects and animations are generally predicated on a somewhat distanced position of spectatorship, interactive effects often foreground auditory effects over visual ones. Most games have a soundtrack that plays repeatedly in the background and is rarely noted by a player, but a soundtrack is different from sound as a special effect. A sound *effect* is a result of a particular action, and when initiated by the player, it is often the occasion for delight and repeated activation.

One example of engagement with an interactive special effect shows an eleven-year-old, Dan, building a city with an undergraduate who is an expert at the game. As Dan is playing with the budget window, he discovers that increasing taxes causes the Sim citizens to boo and lowering them causes them to cheer. He takes some time out from administering the city to play with this auditory effect (line 1, 5), before the undergraduate calls him back to his Sim mayor subjectivity (lines 4, 6).

D = Dan
UG = Undergraduate
square brackets signify overlapping talk

1   D: (Starts bumping up the property tax, big grin.)
2   UG: What are you doing? No, no, no.
3   D: No, I just [want to see . . .]
4   UG: [Now,] now—[listen].
5   D: (Bumps down the property taxes, making the citizens cheer.) [Yeeeee] eaaah. [I just want to make them happy.]
6   UG: [The best way to make money]—You want to increase your population, right? So you lay down the green, right? So if you put all, make all this all green, then, ahh, your population will increase, and then you could raise taxes, and then you could get up to your five thousand mark.
7   D: Ohh, OK. (Closes budget window.)

Dan's apparent pleasure in this interaction can be understood as a kind of computer holding power (Turkle 1984) based on the logic of the interactive special effect. The combination of direct interactional engagement with the machine and a unique responsiveness creates a brief but tight interactional coupling between Dan and *SimCity 2000*. This kind of interactional pleasure occurred numerous times during my observations of kids' play at the 5thD, but was initiated only in the children who were controlling the mouse. Although surface readings of the interface can invite collaborative interpretations, as in the sequence with Jimmy and Holly, the interactive special effect is somewhat antisocial, relying on a tight interactional coupling of human and machine, often at the expense of other interlocutors. As in most examples of this sort, the undergraduate calls Dan back to the more functional and progress-oriented accountabilities of game play. This undergraduate is more heavy-handed than Holly in the previous example, insisting that the kid pay attention: "No, no, no . . . now, now, listen. . . . The best way to make money—you want to increase your population, right?" (lines 2, 4, 6).

An instance of play with the game *DinoPark Tycoon* also exhibits similar dynamics involving interactional special effects. At the "Dino Diner,"

Figure 3.11
The Dino Diner in *DinoPark Tycoon*. Screen shot reproduced with permission from
The Learning Company, Inc.

the player is able to purchase items from a menu as feed for the park's dino-
saurs. One of the features of this screen is that a fly buzzes around the menu,
and if it lands on the menu, and a page is turned, the fly is crushed, emitting
a squishing sound, and the player, upon flipping back to the page, sees a
gooey smudge (figure 3.11). In instance after instance of play with *DinoPark
Tycoon*, kids play repeatedly with this game feature. For example, almost
every time a player named Ian visited the Dino Diner, he spent time smash-
ing flies:

I = Ian
A = Adult

1   I: (Turns a page, and squishing sound results.) Yeah, I just crunched some
         more. Yeah, look at all them. They're so dead (laughing). This is rad. Oh, come
         on, fly, I want you to come down here. Come down here, puppy. Come to
         Papa. Crunch! (Turns page and laughs.)
2   A: That's nasty.
     I: (Turns page.) Crunchie, crunchie, crunchie. (Turns page.) I crunched him!
         I crunched him! (Turns page.) I'm so mean. I want to go check out my dino.
         (Leaves Dino Diner.)

As with Dean, this interaction is relatively brief and clearly peripheral to
the game's primary goals, which are to build and administer the virtual

theme park. Ian takes some time out to enjoy the interactional special effect (lines 1, 2), but returns fairly quickly to the task at hand, checking up on his dinosaur (line 3).

In the discussion of *MSBEHB*, I noted how the auditory effects were one attractive aspect of the game for the kids. One area of *MSBEHB*, involving a simple painting program, is also particularly notable as an embodiment of the logic of the interactive special effect. Clicking on the drawing pad of one of the characters calls forth a screen with a canvas and various tools shaped like body parts. After selecting a body part, the player can squirt, splat, or stamp blobs and shapes onto the canvas, accompanied by gross bodily noises appropriate to the body part. Often to the accompanying undergraduate's dismay, kids at 5thD would spend excruciatingly long minutes repeatedly squirting juices from the stomach or emitting a cacophony of farting noises from the tongue tool. One undergraduate noted after playing with a group of girls: "Each different shape or design made it's own unique sound. I think the kids get a much better kick out of the sound than anything else. And they would laugh and laugh when they found the sound they liked best." We captured an example of such an instance of play in our video record.

R = Ralph
UG = Undergraduate

1   R: Look, I could pick any one of these. This one. (Selects an organ.)
2   UG: What's that stuff right there?
3   R: (Squirts juice out of an organ.) I don't know. Squeezing all the juice out of him.
4   UG: Lovely. Now what happens if you grab that one?
5   R: Big one, the big one. Blood, brain. (Continues to select organs and make blobs and squishing and squeaking noises.)
6   UG: Oh, you can do it on here? Oh, it does a print of what the brain looks like.
7   R: Oh man.
8   UG: You can do the mouth, you haven't done that one.
9   R: Spitting, it's spitting.
10 UG: I know there's more down this way too. Skin, oh that just changes the color of it.
11 R: Yup. Do you want to see the nose?
12 UG: Nose. I don't know.
13 R: Gross.
14 UG: Oh, gross.
15 R: Boogers, eww.

(Continues through each organ in a similar manner.)

16 UG: Your tongue. Oh wow.

17 R: (Creates long, drawn-out farting noises.) Ewww!!! (Pushes repeatedly on a squeaking, blapping organ.) OHHHHH!!

18 UG: Wow.

19 UG: What else is there that you could do?

20 R: Nothing.

21 UG: Is that the last one?

22 R: Yeah.

23 UG: Are you sure?

24 R: Yeah.

25 UG: How do you know that?

26 R: (Goes back to the farting noise and hits it repeatedly.) It's my favorite. The tongue. Watch.

27 UG: Are there any more?

28 R: Oh yeah. The finger. (Clicks repeatedly, making more gross noises.)

29 UG: Eww.

30 R: Ewww. Look.

31 UG: Are there any more after the finger? Let's see. Muscles. Whoa.

32 R: I want to go to the finger again.

33 UG: Why don't we go back and explore the body?

34 R: OK. I know why. I know why. 'Cause you didn't like the sounds.

35 UG: No, the sounds were great.

36 R: I don't want to play anymore.

Ralph gleefully engages in this extended sequence of play with interactive special effects, but the undergraduate only tolerates it, and she eventually suggests that they return to the main areas of the game (line 33). Ralph then suggests that she is discouraging him from playing with gross sounds (line 34) and decides that he wants to stop playing rather than return to the more educational sections of the game (line 36). The undergraduate is remarkably patient through a very extended sequence of play with each drawing tool, suggesting on various occasions that Ralph try one or another tool. Yet the boy still insists that he knows why she suggests that he move on, "'Cause you didn't like the sounds." In this case, the boy is more active than the undergraduate in constructing the opposition between the adult stance and the kid stance with respect to the gross special effects.

Interactional special effects in computer software are similar to the manipulations possible with physical materials such as clay and finger paints, but are mediated by a computational artifact that uniquely amplifies and embellishes the user's actions. Like the visual special effects described earlier, these interactional and auditory effects are not part of a broader game goal structure, but are rather engaged in for momentary and aesthetic pleasure. They are not the dominant modes of engagement in play in children's software, but they are small, ongoing breaks in the multimedia titles' narrative trajectories. They are also sites of micropolitical resistance to the progress-oriented goals and adult values that seek to limit violent and grotesque spectacles in an educational setting such as the 5thD. I return to this dynamic at the conclusion of this chapter.

Mobilizing Fun: The Micropolitics of Pleasure

Engagements with special effects are not merely an atomized individual process. They are part of the economy of "cool," a central source of cultural capital in kids' peer relations. Spectacle and "fun" are mobilized as a device to enlist other kids and to demonstrate style and status, as well as to demarcate a kid-centered space that is opposed to adults' progress goals. A search for all instances of the word *fun* in our video transcript record revealed many instances of the word *funny,* but relatively few instances of children describing something as "fun." More often adults used the term in querying whether a child was engaged: "Are you having fun?" and "Is this a fun game?"

In undergraduate field notes, *fun* most often appears as a descriptor of play characterized by high energy and committed engagement. "This was a fun day. Everyone was really into the new game." "She told me she wanted to play more and so did I. It was a fun game." "She found that to be something fun because she was really excited." "What we ended up doing was just working as long as we could on the same game in an effort to finish it. It was easy to do this because [he] had fun with this game in particular." Less frequently, *fun* gets used as a descriptor of activity that is opposed to the site's progress goals. For example, in one field note, an undergraduate comments, "Today the kids got to play games for fun," meaning that they did not have to play according to the maze activity system that day. As indicated in their notes, undergraduates sometimes saw kids' singular commitment to fun as a problem. An undergraduate writes critically of one boy who would not share the mouse. "For kids, their own fun tends to come first before others, and helping out others may not be their idea of fun." "I reminded Ian that it is never any fun when one person is playing a game and everyone else has to watch that person have all the fun."

In the small number of instances in our transcripts where children use the term *fun*, it describes activity that was spectacular in nature and nonfunctional. In one instance of play, one girl has taken a liking to *SimTown*, a more child-oriented version of *SimCity*. Another child appears, and a discussion of taste and style ensues. "I like *SimCity* better," declares the boy. The girl counters by saying, "No, this is much funner, you don't have, you don't have a debt or anything"—noting her appreciation of the more open-ended structure and lack of financial accountability in *SimTown*. The debate does not end here, though. "So. Who cares? Debts are cool," states the boy. He decides to test the coolness quotient, however, and pushes the girl to check out if there is a disasters function on the game. "Oh, disasters? What are the disasters?" Like "cool" effects that code for spectacle, *fun* in this exchange describes playful, noninstrumental game features. The same girl describes the function where a player can trigger cute animations in the buildings as "fun spots" in contrast to the functional roles of buildings in structuring the town. Another instance of talk between some kids exhibits a similar dynamic. Three kids (denoted by A, B, and C in the script) are playing *SimTown* together for some time, and the one occurrence of the term *fun* happens when they are trying to squash some people riding bikes.

1   A: Oh, look people are coming out of their houses and riding their bikes.
2   B: I'm going to squash them.
3   A: I know.
4   C: Oh, no.
5   A: This looks like a fun game.
6   C: I know.

The use of the term *fun* between kids is a device to enlist peers into a space of shared pleasures. "This is the fun part, look." By contrast, fun can create a boundary between adult goals and child pleasures. In two instances, kids authenticated their activity by describing what they were doing as "having fun," appealing to adults to let them engage in activity that is not progress oriented. Activity that is not directed toward a particular adult's goals are described as "just for fun" even though that same activity might in other contexts be an achievement-oriented task. Children can be politically savvy about the uses of fun, realizing that it is a legitimate form of child-identified activity that can provide a space of self-determination. In one example, when Chris is playing *Dr. Brain,* the site director stops by, and there is a discussion of whether this game counts in the maze that structures the children's movement through the club activity system. "Well, you may never become a Wizard's Assistant," the director warns. "I don't care," Chris replies as he continues playing. "I am just going to have fun."

In another instance, Ian is playing a game of Solitaire, and a video ethnographer comes to set up on the machine Ian is using. The ethnographer works to get Ian's attention and eventually moves to the machine to quit the program.

I = Ian
E = Ethnographer

1   I: Man . . .
2   E: Yo, Ian, we're taping on this machine, so get somebody to play with. Hello, are you listening to me?
3   I: What?
4   E: I'm going to exit you.
5   I: Why? I'm having fun.
6   E: Because this is . . . we're spending money to videotape at this machine, and unless you want to help somebody play a game, then get off of it.
7   I: I'll help them play this.
8   E: No, this isn't a game in the maze.
9   I: So? I can make it be one.
10   E: Who says?

(Ethnographer takes the mouse and exits Solitaire.)

In this instance, Ian fails to claim a space of indulgence and is made to participate in the club's sanctioned activities.

These two examples are suggestive of kids' awareness of fun as a legitimate site of resistance to certain adults goals. The 5thD's adult narratives also reinforce this sense of fun as a site of authentic and natural childlike agency that can be harnessed to promote engagement, but that can also overpower the play setting if not channeled in a prosocial way. In the 5thD's micropolitics, just as in the packaging of entertainment for children, fun appears as a cultural fact that is part of the contemporary construction of childhood, a political tool that gets mobilized to appeal to adults to suspend their own agendas. This political role of fun is also closely tied to the vernaculars of hedonistic, repetitive forms of play and a child-centered visual culture.

Disobedience, Disasters, and Action Entertainment

In the 5thD, an orientation to entertainment (i.e., "fun") is actively encouraged, but ultimately in the service of a progressive educational project. Children mobilize fun as a way of indicating authentic engagement, and fun is celebrated in the 5thD to the extent that it happens in the context of a prosocial learning task. "Entertainment" is clearly not a monolithic category within commercial media forms. Although the 5thD project legitimizes some entertainment idioms, it explicitly excludes action-gaming idioms as too patently noneducational. As noted in the previous chapter, action-entertainment idioms are constantly lurking in the ambient culture that kids participate in. These cultural elements are largely repressed in the 5thD through the selection of nonviolent games and persistent adult surveillance, but they are still present. Due to their illegitimate status in the 5thD, they become a resource for subverting dominant (educational) codes in this local context. The case of Ian and *SimCity 2000* makes clear this relational dynamic between educational and entertainment idioms in ways that are particular to the 5thD, but that point to the pervasive fractures in U.S. culture between "wholesome" educational norms and violent entertainment idioms. These fractures are hints regarding the pervasiveness of action entertainment in peer dynamics, particularly of boys. I present these final examples of engagement with the entertainment genre as a special case of engagement with spectacle, a much larger topic that my work in the 5thD only partially touches on.

One day of Ian's play with *SimCity 2000,* captured on video, is a rare case in which action entertainment appears as a social resource in the 5thD, and it enables us to see the tensions around this cultural domain as it appears in an informal learning setting of this sort. This instance of Ian's play forms a more focused case study in chapter 4, but here I introduce a portion of it to illustrate the social role of violent and destructive entertainment. The scene opens with Ian sitting in front of the computer, interacting with a well-developed city marked by an enormous airport and waterfalls stacked in a pyramid formation. Another boy is sitting next to him, observing his play and making occasional suggestions, and Ian also has an audience of other club participants, including the video ethnographer, undergraduates, and other kids and adults who are walking in and out of the scene. Ian busily makes a railroad system, water pipes, buildings, and a power plant, and he worries about such things as whether his people are getting enough water and whether power plants need to be replaced. The site director appears and tries to get Ian to teach others how to play (line 1), but Ian deftly deflects this accountability to the club norm of collaborative learning and receives another kid's support (line 2).

I = Ian
M = Mark (a younger boy)
SD = site director

1   SD: Because you're not going to be sitting here all day just doing it by yourself. So other people watch you. It's not fair to other people.
2   M: No, we, we, we, we're not supposed to be able to play. We're not supposed to play.
3   SD: Why aren't you supposed to play?
4   I: They're not.
5   M: If you're not a Young Wizard's [Assistant], you can't play this.
6   SD: But if you're a Young Wizard's Assistant, and you're not teaching anybody else the game, then you can't play it either.
7   M: He's teaching me.
8   I: (Unintelligible) said I could.
9   SD: OK, good, all right, check it out then.
10   I: Anybody ask me any questions.

Ian's tactic is momentarily successful; he passes as a teacher and resumes his game play. After about twenty minutes, however, the site director interrupts him again and asks him to teach a new undergraduate how to play the game. "I'm not kidding, either," the director stresses. "Her grade depends on what you teach her, so she'd better do a good job, OK?" After a few moments, Mark suggests, "Show her a disaster. Do an airplane crash." Ian responds with enthusiasm, saves his city, and announces, "Ha ha ha, disaster time!!"

In this sequence of activity, Ian finds himself in the center of a series of interventions and a great deal of social attention, positioned as an expert and asked to teach both an undergraduate and a large audience of other kids about the game. The video ethnographer and the site director have already intervened a number of times to orient him to his community role as game expert and teacher. His companion is the first to suggest doing a disaster, and Ian takes it up with a characteristic virtuosity and antiauthoritarianism. Disaster time involves an escalating series of special effects in which the city is first invaded by a space alien, then flooded, set on fire, and subjected to an earthquake and plane crashes. The undergraduate remains pleasant and amused. The video ethnographer, a veteran at the club, is the first to intervene, addressing the undergraduate first. "So, have you figured out how to play?" And then she turns to Ian. "Remember Ian, that Anne has to . . . , Ian?!" The video ethnographer and undergraduate's protests punctuate this instance of play, and though they do not specifically deny the appeal of destruction, they are clearly trying to redirect the activity. They are overpowered as Mark cheers Ian on, and the two boys delight in the spectacles of destruction. "Do another airplane crash!" "Destroy it." Yet another boy joins the spectacle. "Please do a fire engine." "Put more fire. Fire's cool." "Just burn it all. Burn it. Burn it. Just burn it. Burn it. Burn it. You need more fire, more fire." The site director appears again. "Is he teaching you how to be a constructive citizen?" he jokes. "Another five minutes, and then put Anne on and see what she can do." "Do riots," the third boy continues, not responding to the director's comment.

After the city is in flames, Ian begins to build large buildings within burning areas to induce more and more spectacular explosions. He turns from blowing up the most expensive of the possible buildings to blowing up colleges, fusion plants, gas power plants, and microwave power plants. His final achievement is to blow up a row of fusion plants lined up in domino formation (figure 3.12). "Ian, time, put Anne in there," insists the site director at the conclusion of this performance. "He's into mass destruction at the moment," says Anne, worried. The director assures her, "Yeah, but these guys know a lot about the game." Then he turns to Ian. "I don't want to turn the machine off on you. Be nice to Anne and give her a turn." That is enough of a credible threat for Ian to start a new city for Anne. Ian's subsequent acts of constructing a city are the case study of chapter 4. Here, though, I am focusing on the role of action-entertainment idioms in enlisting an audience of other boys and the role of computational media in enabling a virtuosity of the spectacular in a player's hands. The adults at the club were in the difficult position of trying to validate Ian's technical knowledge, but not wanting the destructive scenario to continue. Ian was quite aware of the boundaries of participation in the 5thD, and he played to his moment in the spotlight until he was on the verge of disciplinary action. Far from being regressive and antisocial, he was engaging in a process of enlisting a large and engaged audience in a shared performance.

Figure 3.12
Blowing up fusion plants in *SimCity 2000*. Screen shot reproduced with permission from Electronic Arts Inc. © 1993–1994 Electronic Arts Inc. SimCity 2000 and SimCity are trademarks or registered trademarks of Electronic Arts Inc. in the United States and/or other countries. All rights reserved.

Ian's instance of play was an unusual one in the 5thD in terms of the prominence of its destructive scenario, but action-entertainment idioms are often latent in play with computer games. On one of my field visits to the Boys and Girls Club during the summer when the 5thD was not in session, I set up a machine with *SimCity 2000* and invited some kids to play on it. "Do you know *SimCity*?" I asked a boy whom I hadn't seen before, maybe seven years old or so. "Yeah!" he declared. "That's the game where you blow up the cities!" Nonplussed, I nonetheless asked him if he would like to play and whether he knew how to, and he answered yes on both counts. As he hunted around the toolbar to perform a certain operation, it became clear that he knew little about how the game worked. Finally, he asked, agitated, "Where are the guns?!" With the help of another boy, he succeeded in setting the city on fire and shortly thereafter left that computer to play a different game. Ian's orientation to the software's action potential was thus clearly not unique. In many other instances of play documented in the undergraduate field notes, boys used the disasters function to destroy their cities after they ran into fiscal trouble.

In another instance of play we recorded in the 5thD, Jimmy is patiently building up his city. One of his bridges explodes because he hasn't allocated sufficient maintenance funds. "I broke it again! My bridge!" he exclaims in distress. Another boy, observing, responds, "Cool. Save it and restart." "And do what?" asks Jimmy, apparently perplexed by the suggestion. The other boy continues, his voice rising: "I love doing that. I love just saving a city and then, just *destroy it!* Every disaster!!" Another child, just checking into the scene, pipes in, "I know! Especially the monsters. Those are cool." "The monster's different on this one," declares the first boy, in the know regarding the new version of *SimCity 2000*. "It is?" "Yeah," he continues, "You've never seen the monster of *SimCity 2000*?" (figure 3.13). The space alien is a peripheral, but unmistakable nod to fantasy worlds continuously reinscribed in computer games ever since the hit game *Space Invaders*. In this brief interactional moment, Jimmy has been educated regarding the not so obvious citational links between

Figure 3.13
The monster in *SimCity 2000*. Screen shot reproduced with permission from Electronic Arts, Inc. © 1993–1994 Electronic Arts Inc. SimCity 200 and SimCity are trademarks or registered trademarks of Electronic Arts Inc. in the United States and/ or other countries. All rights reserved.

SimCity 2000 and action-gaming idioms, where such disparate cultural elements as *SimCity 2000, Space Invaders,* monster movies, and "coolness" are tied together.

My observations of kids, usually boys, who are immersed in computer gaming and action-oriented media is that they index a wealth of cultural content that weaves together representations in film, television, and interactive media. As Marsha Kinder (1991) explains in *Playing with Power,* games such as *Mario Brothers* and *Mortal Kombat* have made their way into movies, and many moves have made their way into interactive media. Action-gaming idioms are clearly the cultural domain that has the most extensive set of referents in interactive media, and kids adeptly interpret, critique, and make distinctions between different aspects of the action-gaming corpus. Even in a context such as the 5thD, which has been designed specifically to exclude violent media, these intertextual citations are irrepressible, an always available source of cultural capital for kids.

Conclusions

If we resist the impulse to call the engagement with action entertainment antisocial, then we can begin to query a certain social paradox. Although competitive achievement that individuates learning and produces class distinction is considered prosocial and developmentally correct, hedonistic play that creates peer solidarity in relation to consumer culture is considered antisocial and regressive, an attention deficit to the progress goals of certain authoritative institutions. Brian Sutton-Smith describes this tension in terms of private and public transcripts of childhood: "The adult public transcript is to make children progress, the adult private transcript is to deny their sexual and aggressive impulses; the child public transcript is to be successful as family members and school children, and their private transcript is their play life, in which they can express both their hidden identity and their resentment of being a captive population" (1997, 123).

But this argument is only one take on the complex sociocultural contestations that children navigate in their everyday play, particularly in relation to an increasingly powerful and institutionalized entertainment industry that produces and inflects our ideas of childhood. The institution of education and the ideology of individual merit are produced in opposition and constant negotiation with their antithesis—"mass" accessible popular culture. Instances of children's play have shown that popular culture, far from being an undifferentiated field of cheaply accessible and passive thrills, is a site of virtuosity, connoisseurship, and status negotiation among children as well as between children and adults. In children's everyday lives, what cultural forms get played out are a contingent effect of local micropolitics, where pop culture identification confers status in children's hierarchies and "fun" gets mobilized vis-à-vis adults as an authenticating trope of a "natural," childlike pleasure principal. This configuration of relationships is not a simple story of adult repression of authentic childhood impulses, but rather a distributed social field that produces the opposition between childhood pleasure and adult achievement norms as one cultural effect. In a climate that increasingly values childhood as a romantic and privileged sphere, adults who discover their inner children and indulge their own children are on the progressive cutting edge of parenting, a trend that industrialists are quick to exploit in their marketing aimed at the parental pocketbook. Children's entertainment industries, new technologies, and the 5thD's practices are part of the discursive productions of these oppositions and changing notions of childhood.

The spectacular dimensions of new media deserve special mention. The atomized consciousness of a player engaging with a special effect is a small moment attached to a large sociotechnical apparatus. Whether in movies or computer games, special effects drive budgets and bring in audiences. This relationship is indicative of a particular kind of industry maturation, where a growing consumer base supports larger production budgets, but also increases investor risk, driving the push toward sure-hit products, sequels, formulaic content, and guaranteed crowd pleasers. Special effects also weed out independent developers who don't have the budgets to compete in production value in similar genres. As the children's software industry matured, the family entertainment genre shifted from playing a more hybrid role in bridging entertainment and learning goals to playing a role more closely aligned to TV culture than school culture. Although a title such as MSBEHB still appealed to educational content, later titles in the genre abandoned these educational cultural markers even for software oriented to young children.

Entertainment industries participate in the production of genres that are packaged and hardened into certain formulas that kids recognize and identify with as a liberating, authentic kids' culture. Just as educational content has been commodified as curricular coverage and achievement anxiety, entertainment genres are packaged into easily reproducible formulas, vernaculars of children's popular culture. In the titles I reviewed in this chapter, these formulas appeared as gross bodily noises, explosions, visual hyperbole, and, increasingly, established licensed characters. This "junk culture" is a particular vernacular that cross-cuts media and commodity types, making its way into snack foods, television, movies, school supplies, and interactive multimedia. Just as this junk culture is a site of opposition between adults and kids, entertainment elements in children's software become opportunities for kids to resist adult learning goals in the 5thD. In choosing to conform to the entertainment genre, software developers make political decisions about where they choose to align themselves in this power play.

The idiom of children's entertainment finds full expression in action media such as video games, movies, and television, which are consumed in the home among peers. My work in the 5thD has allowed me only a small glimpse into these domains. In children's software and even in educational contexts such as the 5thD, the entertainment genre is a constant though often latent presence, evidenced in the appreciative exclamations of "cool," "awesome," "Eeeew!" and devilish cackles of delight that punctuate game play. These remarks and moments of play are constructed in interaction with adult efforts both to identify with and to resist these idioms, which are tied to a mushrooming children's media system that embodies the growing strength of the entertainment genre. As the pleasure principal becomes coded and embodied as the domain of an authentic and natural childhood, (boring, dry, dusty) education gets framed in an act of symbolic and social violence as going against this childhood "nature." The cultural assumption embedded in this genre framing is that learning needs to be fun to be authentic because fun is what kids do best. Edutainment's platitude that "learning can be fun" reconstitutes this structural opposition between learning and play, while simultaneously working to deconstruct it.

4 Construction

Seymour Papert is one of the best-known spokespersons for the use of computers in education, specifically the use of computer programming as an educational tool. Although part of a shared intellectual community and discursive tradition as McCormick and many other educational software developers, Papert is distinctive in his promotion of programming as a key educational and developmental goal. In contrast to the focus on content and curriculum that characterizes most educational software, for Papert, content is secondary to what he calls "technological fluency," or the ability to perform the new forms of literacy enabled by computer technology. Published in 1980, his book *Mindstorms* describes the LOGO programming language designed for children, arguing against the drill-and-practice orientation of computer-based instruction that was dominant at the time: "In most contemporary educational situations where children come into contact with computers the computer is used to put children through their paces, to provide feedback and to dispense information. The computer programming the child. In the LOGO environment the relationship is reversed: The child, even at preschool ages, is in control: The child programs the computer" (19).

The focus of his criticism differs from the position that most advocates of learning software take. He is objecting to the drill-and-practice approach not because it is, in McCormick's terms, a "dry and dusty" form of learning, but because it does not allow the child the subject position of agency and authorship. Even in his more recent publications, in an era when computers are becoming widely available to children, Papert is critical of much of the edutainment software on the market. "Disguising flash cards as a game introduces an element of deception that undermines two fundamental educational principles. First, learning works best when the learner is a willing and conscious participant. Second, deception and dishonesty in a teaching process make a mockery of the idea that schools should develop moral values as well as knowledge of math or history" (1996, 19). "The dominant trend in educational software is following a path that bothers me," he asserts. "The mildest criticism that I can make of it is that it panders to popular prejudices about what is 'educational.' The more severe criticism is that most educational software powerfully reinforces the poorest sides of pre-computer education while losing the opportunity to powerfully strengthen the best sides" (1996, 37).

"What I see as the real contribution of digital media to education," he says, "is a flexibility that could allow every individual to find personal paths to learning" (1996, 16). His list of recommended software strongly favors programs that lean toward user authoring—what he calls *constructionist software*—rather than the adventure- and quiz-type formats dominant in current educational software. For example, he suggests the paint program *KidPix* and simulation games such as *SimCity* and *SimTower* (1996, 209), and he has been working since the late 1970s on designing, promoting, and upgrading the LOGO programming language, now a multimedia authoring tool called *MicroWorlds*.

In the first section of this chapter, I review the history and cultural context of construction software in relation to "hard mastery" of computers associated with hacking subcultures. I then describe how the authoring genre of children's software is associated with these kinds of uses of computers and how this association manifests itself in the title *SimCity 2000*. The last half of the chapter describes how *SimCity 2000* and a related title, *SimTower*, were used in the 5thD to create personalized virtual worlds.

Hacking, Making, and Geek Identity

Drawing from Piaget, Papert describes his orientation as "constructivist," based on "a model of children as builders of their own intellectual structures." He sees the computer as providing new sorts of materials for the "child as builder," making abstract mathematical concepts concrete and manipulable (1980, 7). Papert's metaphors of construction put him in the same intellectual trajectory as those educationally minded toymakers who brought us building blocks and erector sets, a trajectory set along a child-centered philosophy of learning that departs from the behaviorist model.

The orientation toward computing as an authoring tool, a device to actual-
ize individual agency, can also be located within an ethic of self-authoring
computer programming and use that has persisted since the earliest days
of personal computing.

In the popular history of personal computing *Hackers: Heroes of the
Computer Revolution,* journalist Steven Levy (1994) describes "the hacker
ethic" that emerged from a group of computer enthusiasts at MIT in the
early 1960s around the first interactive computers that allowed a programmer to get immediate feedback from a computer terminal. "Access to
computers—and anything which might teach you about the way the world
works—should be unlimited and total. Always yield to the Hands-On
Imperative!" was their view (40). Following from this view were such
beliefs as "All information should be free"; "Mistrust Authority—Promote
Decentralization"; "Hackers should be judged by their hacking, not [by]
bogus criteria such as degrees, age, or position"; "You can create art and
beauty on a computer"; and "Computers can change your life for the
better" (40–45). Levy's and my use of the term *hacker* should be distinguished from the subsequent popularization of the term to refer to unlaw-
ful activity and instead reference a more general orientation toward the
computer as a tool of empowerment and discovery. Levy goes on to
chronicle how this orientation toward computing extended beyond what
he calls "the monastery" at MIT in the 1960s as computers became widespread in the form of the "personal computer." In the 1970s, Berkeley,
California, was a key site of grassroots computing, with Ted Nelson's (1974)
hacker cult publication *Computer Lib* and the ongoing meetings of the
Silicon Valley Homebrew Computer Club, both of which encouraged users
to take control of their computers. Rebelling against mainstream corporate
computing as defined by IBM, this orientation toward computing eventually led to Stephen Wozniak and Steve Jobs's development and marketing
of the Apple II in the late 1970s. Personal and hobbyist computing led to
the new fortunes made by game hackers in the 1980s and eventually to
the mainstreaming and commercialization of the PC industry, dissipating
much of the antiauthoritarian, anticapitalist counterculturalism of the
early years (Levy 1994).

In this hacker orientation, the status of the computer as tool for authoring and making are central. Engaging and enlightening content is secondary to the political positioning of mastery and self-authoring; transparent

access to and control of the technical layers of the machine are sought over surface appearances and spectacle. In her study of Internet communities, however, Sherry Turkle describes how a "postmodern aesthetic of simulation," stressing "soft mastery" and manipulation of surface image over "hard" technical mastery, has recently come to dominate computing (1995, 41–49). This shift has happened at the expense of the hobbyist and hacker's earlier rebel orientation. Yet, as Turkle also acknowledges, the hacker ethic nevertheless persisted as a subculture of computing. Lori Kendall, in her study of an online space populated by programmers, discusses "a form of masculinity that is convergent with computer culture, itself a masculine domain" (2002, 73). This form of masculinity is defined in opposition to mainstream notions of masculinity and is centered on a sense of power over technology. The computer's interactive qualities are valued not only for responsively delivering information and feedback, but for enabling users to embody their agency computationally: as Papert states, users' programming the computer rather than the computer's programming them.

One important counterpoint to a univocal celebration of hacker culture comes from feminist analysis. Work on feminism and technology describes the persistent and pervasive cultural bias that equates men with technologies of power and reduces the significance of technologies associated with women. Concluding her feminist analysis of technologies of work, reproduction, home, and space making, Judy Wajcman writes: "The enduring force of the identification between technology and manliness, therefore, is not inherent in biological sex difference. It is rather the result of the historical and cultural construction of gender" (1991, 137). A similar bias operates in the domain of computer use. Although women are associated with routine clerical uses of computing, they are not considered candidates for the "hard mastery" and innovative uses associated with hacking, where the person is in control of deep technical knowledge. Kendall discusses how men in an online group she participated in recognize that they do not conform to hegemonic notions of masculinity, but still reproduce norms of heterosexual masculinity, including the objectification of women. Studies of technology use in classrooms document how the most technically sophisticated boys tend to be from more privileged backgrounds, and boys who identify with more working-class notions of masculinity often shun computers (Holloway and Valentine 2001; Seiter 2005). Although girls are often competent computer users, they tend not to take on the practices or identities associated with boy geek culture (Holloway and Valentine 2001; Livingstone 2002; Seiter 2005; Silverstone, Hirsch, and Morley 1992).

Turkle describes a shift in the mid-1980s away from hard mastery and toward the dominance of "soft" and more women-friendly forms of computer use. In her view, women hackers in the early years felt that they needed to distance themselves from close identification with the computer, even though they were attracted to it (1995, 62). In the recent turn toward more graphical and concrete interfaces, Turkle sees hope for transformation in the relation between women and computers, an emergent "culture of simulation" where "the computer is still a tool but less like a hammer and more like a harpsichord" (63). "A classical modernist vision of computer intelligence has made room for a romantic postmodern one. At this juncture, there is potential for a more welcoming environment for women, humanists, and artists in the technical culture" (62). In the 1990s, we saw a movement toward the design of "girls' games" that aimed to broaden the gamer demographic. Many celebrated the effort to create female-identified genres of gaming. Many feminists, however, critiqued the movement for taking gender difference as a given and for reifying it by making games that are less aggressive, less competitive, and oriented toward girl-friendly themes such as friendships, fashion, storytelling, and popular music (Cassell and Jenkins 1998).

In gaming, as well as in other domains of digital mastery, women and girls have had a role that can't be reduced to a "feminine" stance of "soft" engagement. L. Jean Camp is a participant in the mailing list Systers, an electronic forum started by Anita Borg to link women computer science professionals. Camp writes, "We are geeks, and we are not guys. Not guys, but geeks! How could that be? If it surprises you to learn that more than fifteen hundred feminist geeks are out there, imagine the surprise to each of us" (1996, 114). T. L. Taylor, in critiquing the girls' game movement, argues that rather than focusing on "essential" differences between boys and girls and trying to make "girl-friendly" environments, the field would benefit from paying more attention to the girls who do play. "Current women players are regularly seen as anomalies and not of central research interest. What this means, however, is that we tend to leave their pleasures, their strategies, their networks, their *play* always at the margins" (2008,

56). Taylor's work looks at women in professional gaming, shifting the focus away from the game content to the game activity and the social networks that support game play and gamer identity.

Taking a cue from Taylor, we should recognize how domains of cultural production with historically strong participation by women share affinities with hacker creativity and antiauthoritarianism. Although the hacker ethic is something particular to computational media, it shares commonalities with other kinds of media "hacking" in literature, television, and film. Studies of fan communities, with their appropriation of mainstream narratives through fanzines and remade video, are examples of a similar orientation to other media types, and these communities are domains where women and girls have often been dominant (Baym 2000; Bury 2005; Coppa 2008; Jenkins 1992; Kearney 2006; Penley 1997). Media hacking is a particular twist to the idea of active and oppositional consumption, a resistance to the passivity normally associated with consumption. In her study of different forms of girl-made media, Mary Celeste Kearney notes "a dramatic increase in girl zinesters, girl musicians, and girl web designers." She sees "the increased availability of inexpensive, user-friendly media technologies for amateurs" as possibly the most significant driver of this growth (2006, 2). In other words, girls' have driven the development of certain forms of media-making cultures, and these cultures are increasingly defining the terrain of digital authoring.

Although the identity of a geek, hacker, and gamer still leans toward the marginal masculine identities that Kendall identifies in her work online, we are seeing a broadening of the demographic of digital media makers as digital media production and Internet distribution have become widespread. Practices of game modding, mashup, remix, and everyday forms of media creation are tied to a blossoming do-it-yourself media ethic that is more widespread than the early hacker counterculture (Benkler 2000, 2006; Ito, Baumer, Bittanti, et al. 2009; Jenkins 2006; Lowood 2007; Ondrejka 2007; Russell, Ito, Richmond, et al. 2008). Since in the late 1990s, when digital media authoring tools on consumer PCs became more sophisticated and the Internet went mainstream, media authoring and technology tinkering have gradually become more mainstream. Practices such as creating MySpace profiles and modifying and circulating photos and videos have become pervasive in U.S. youth culture (Ito, Baumer, Bittanti, et al. 2009). Although these more recent trends in digital culture do not have the same subcultural capital as the trend created by early hacker and geek communities, they are evidence of the growing strength of the genre of construction in young people's everyday media use.

As the Internet has become more mainstream, girls have come to dominate certain online spaces. The more "social" uses of online networks—exemplified by social network sites such as MySpace and Facebook, and by the use of chat technologies such as instant messaging—have changed the gender valences of the online world. Girls are now online in numbers that rival the boys, but this increase in their presence has not erased the gender connotations associated with particular technology practices (Mazzarella 2005). Girls' presence online has prompted a discourse of risk, fear, and protectionism that was less present in the days of the geekier male-dominated Internet. Women and girls continue to be framed as the victims rather than the masters of technology practices (Cassell and Cramer 2007; L. Edwards 2005). In gaming, the past decade has seen a similar growth in participation by women and girls. Yet these gains are largely attributed to the growth of "casual" games, such as puzzle games and other Internet games, or games such as the *Sims* that depart from more stereotypically masculine game genres. In their introduction to *Beyond Barbie and Mortal Kombat*, Yasmin B. Kafai, Carrie Heeter, Jill Denner, and Jennifer Y. Sun revisit the state of gender and gaming and argue, "Although the presence of women and girls in a range of game worlds is encouraging, most games continue to replicate and perpetuate the gender stereotypes and inequities found in our society" (2008, xii). They note that women are still radically underrepresented in game design and that discourse of gaming still revolves around simple boy-girl binaries.

The late 1990s, when I was conducting my research in the 5thD, was a period of tremendous flux in the gendering of technology and the spread of digital culture in the everyday lives of diverse young people. Computers were quickly moving away from their geeky and ghettoized origins to a more expansive social palette. One consequence of this spread was girls and women's greater participation in digital and geek culture, though this growth did not necessarily always challenge existing gender distinctions. Gender distinctions are resilient, but it is important to note the ways in which women and girls' participation in media making and social online media has challenged the dominance of boy-geek culture in these domains. When I was conducting my fieldwork, the online world was rapidly transforming into a space where player activism and authoring as well as forms of creativity and remix that were less clearly gendered, were gaining a stronger foothold. The contemporary landscape of online media has evolved well beyond what Papert originally imagined when he was promoting the LOGO programming language, but in many ways the genre of construction that he originally put forth has become culturally dominant in the era of contemporary networked digital media. The computer's ability to respond to its user's agency and vision is what makes digital media so distinctive and characterizes some of the most groundbreaking software titles in children's as well as adult genres. Early software in the authoring and construction genre and kids' uptake of these technologies were indicators of how this genre of participation would grow in tandem with the spread of new forms of digital media production tools and online social exchange.

Budding Hackers: Promoting Technical Literacy and Multimedia Authoring

Papert's work with educational software from the 1970s is an effort in hacker literacy, an interface between a constructivist educational agenda and computer programming and authoring capabilities. The LOGO programming language, developed in that decade and commercialized in the 1980s and later, provided a simplified programming environment that allowed children to see the visual representations of their programming. A child could instruct a "turtle" to move and turn in different directions as well as create drawings and music. Papert's argument was that these programming tasks gave children a set of materials to engage intuitively with basic mathematical concepts. Papert coined the term *constructionism* to refer to a merging of Piagetian constructivism and the image of a construction kit (1993, 142–143). His students and colleagues at MIT have applied this constructionist approach to the design of new construction software and tools such as *Cricket* and *Scratch* as well as learning programs in schools and computer clubhouses (Harel 1991; Hooper 2007; Kafai 1995; Resnick 2006).

Unlike *Oregon Trail, Math Blaster,* and *Reader Rabbit,* LOGO, *MicroWorlds, Scratch,* and *Cricket* are not major commercial successes as consumer software, having the strongest influence in schools. Papert's theoretically motivated claims that programming leads to math and science learning have not been as amenable to packaging and marketing as have been the educational claims by software that clearly represents academic content areas. I include Papert's line of work here not because of these particular software efforts per se, but because of the influence that Papert has had in giving voice to an orientation toward children's computing that extends to their engagements at school as well as in the home. As the constructionist movement matured, the focus shifted from Papert's early orientation toward programming to a more general orientation toward computing and design that puts the child in control of the learning experience. For example, Idit Harel (1991) and Yasmin Kafai (1995) created school-based programs that centered on the design and sharing of software and computer games. Papert recommends Web authoring projects, drawing with *KidPix,* and play with simulation games in addition to use of his own programming and multimedia authoring software *MicroWorlds* (Papert 1996). In the broader genre of children's authoring-oriented software, the two types of software that have proven most commercially viable are not programming tools, but kid-oriented graphics programs such as *KidPix* and *Print Artist* and simulation games, most notably *SimCity* and *The Sims.*

Unlike edutainment software with its age-graded framework, and unlike children's entertainment, which is coded in opposition to adult culture, authoring programs do not posit a sharp break between the adult market and the children's market. Although the MIT group has created software designed specifically for children, most authoring software has been designed as adult-oriented tools that are adapted downward for children. Programs such as *KidPix* have simplified controls, sound effects, and kid-oriented graphics, but they are functionally quite similar to adult applications. In the case of *SimCity,* a game designed for adults was taken up by the kids and the educational market. Parents I spoke to described to me how their children would use business-oriented authoring tools such as PowerPoint to design cards and pictures for their friends and family. At the time I was observing, the 5thD used standard word-processing tools such as Microsoft Word for writing tasks at the site. This particular genre of children's software was not about creating a new category, but about making existing authoring and construction tools accessible to children. This genre of software puts adultlike agency in the hands of children, moving them away from a more childlike receptive stance.

In my interview with Will Wright in 1999, the creator of the original *SimCity*, he described how he did not target children in the design and initial marketing of the game: "I didn't realize that people would take it that seriously. I thought it might have some limited appeal to city planning types, but for the most part, they didn't play games. . . . We tried to make a game that we would like to play, a little more thoughtful, a little more interesting."

His market is age blind. "So the fact that there are seven-year-olds playing *SimCity*, that's great. That's exactly the way that I would want to see the kids' market, as opposed to something with big brightly colored buttons and some cute fuzzy character." He distinguished between his software and games that have competitive goal orientations. "I like games, but I'm kind of uncomfortable with that term for some of the stuff we do, because I think some of the stuff we do is more of a toy than a game." He sees *SimCity* as more of an open-ended tool, linking it to a hobbyist's pleasure in creating, constructing, and designing something unique and personalized. His work melds the constructivist orientation toward tool development with the visual and interactive appeal of computer games.

A toy, I think, is quite a bit more open-ended. A toy you can actually use for a lot of different games. You can come up with your own rules. You give someone a ball, and a ball is not a game, but you can play a lot of games with a ball because it is so open-ended. In some sense, the stuff that I really enjoy doing, I would say with the *SimCity*-type stuff, is closer to a hobby than a game. I think of it in those terms, again that kind of construction and creativity part. I really like things where I can build something, design something. I want to be able to do something in the game that nobody else has ever done with that game. So no two cities are alike. When you play *SimCity*, they may be similar, but no two are identical. There's something really cool about that.

In contrast to McCormick or Davidson's pedagogical stance, Wright argues for the value of *SimCity* not in conveying particular forms of content, in "teaching" urban planning, but in providing an open-ended set of tools that allows intellectual exploration and creative production of a simulated object. He also does not posit a sharp break between his own subjectivity as a "producer" and a player's subjectivity. As he spoke during the interview, he alternated between identifying as a player and identifying as a software developer. This is not to say that content and his role as designer do not matter to Wright; he sees software as delivering a certain message and is a meticulous researcher in the domains he delves into. He described some of his more science-oriented games as having a Carl Saganesque quality of scientific popularization, providing a dynamic model of scientific theories such as the Gaia hypothesis *(SimEarth)* or emergent behavior *(SimAnt)*. Yet his research into content domains is not an effort to represent authoritative knowledge realistically, but rather an exercise in experimentation and exploration of an imagined and hypothesized space.

I think if we tried to make it realistic, we would be doing something that we wouldn't want to do. Many people come to us and say, "You should do the professional version." That really scares me because I know how pathetic the simulations are, really, compared to reality. The last thing that I want people to come away with is that we're on the verge of being able to simulate the way that a city really develops, because we're not.

Other games have followed in *SimCity*'s footsteps, providing various kinds of simulation authoring capabilities. In the children's market, there are *DinoPark Tycoon, Rollerpark Tycoon, DroidWorks,* and a host of other "Sim" imitators.

Although the construction genre has had a relatively limited scope in the market of consumer products for educating young children, its reach in the culture at large has expanded massively in the years since I conducted my research in the 1990s. Software in this genre has been successfully linked to the toy industry. The MIT LOGO group partnered with LEGO in creating *LEGO Mindstorms,* a kit that allows kids to program and control LEGO vehicles. Another major license entered this genre with *Barbie Fashion Designer,* a successful Mattel product that allows users to design fashions and print them out onto fabrics to create Barbie outfits. This genre's more toylike orientation makes these tie-ins a natural fit. More recently, the gaming industry has seen an overall shift toward participation genres with a more construction-oriented bent. For young children, games such as *Pokemon* and *Yu-Gi-Oh* popularized a more player-centered orientation, where kids remix, trade, and customize their game play within a social context (Ito 2006b, 2007; Tobin 2004). Mainstream gaming has also seen a decisive shift toward player-driven content creation and active participation. Some indicators of this shift include the growth of multiplayer online worlds and games (Boellstorff 2008; Ondrejka 2007; Taylor 2006), machinima ("machine cinema") (Lowood 2007), alternate-reality games (McGonigal 2006), and the vast economy of "gamer capital" that circulates on the Internet in the form of commentary, reviews, tips, and cheats (Consolvo 2007). Although these kinds of activities may seem a far cry from kids' programming turtles in LOGO, they are tied into a similar genre of participation that centers on player design and construction of meaning.

Packaging Power

Barbie and LEGO are particularly appropriate licenses for the construction genre because they capitalize on existing brands tied to open-ended toy play. Other titles have had a harder time entering this niche because the construction genre is more difficult to position and market than formulas more closely tied to educational or entertainment content. For example, Lucas Learning, the children's software arm of Lucas Film, released *DroidWorks, Pit Droids,* and *The Gungan Frontier,* three construction-type titles, as their first releases. They were initially marketed and packaged as entertainment titles, attempting to capitalize on the *Star Wars* license and mobilize it for the children's market and educational goals. Ads ran in both gaming magazines and *Family PC.* With these titles, Lucas Learning was competing with mainstream gaming in terms of production value and market positioning. It utilized expensive, customized three-dimensional game engines and an original concept. Eventually, though, the orientation of the company's products and the products' box design shifted from the more mainstream entertainment genre to the family edutainment genre. The original packaging of *Gungan Frontier* framed the product within the adult entertainment genre—gold fonts on a darkly ominous background and a scary, frowning character. The box was later updated to feature the same character, but with a smile rather than a frown and surrounded by bright yellow banners and cute purple lettering. In subsequent products, Lucas Learning lowered the game's target ages to less than ten years and entered the more established children's entertainment genre. All three Lucas Learning titles embed quasi-educational tasks in a fantasy scenario featuring the company's more kid-friendly characters such as Yoda and the young Luke Skywalker. The boxes for these titles were designed for shelving in the kids' corner rather than in the gaming section, and they feature wide-eyed, cute characters that fit right in with *Reader Rabbit* and Disney titles.

In a 1999 interview, Jon Blossom, who worked with Lucas Learning project leader Colette Michaud on *DroidWorks,* described some of the difficulties that he had in positioning his product given existing genres: "Lucas Learning is really interesting in that we're trying to straddle that line between education and entertainment. It's been a really, really hard line to travel because it's not really clear how you market that. It's not really clear how you explain it to people." He contrasted his work with the standard curricular products, the "two plus two kinds of products that we don't think are fun," and commented on the difficulties of marketing learning as a genuinely attractive and fun activity for older kids: "It's hard to convince parents that our games are educational, and it's also generally hard to sell to a preteen, particularly someone who's ten or eleven, trying to be independent, thinks learning is dumb, and doesn't like going to school, and here's this box that says 'Lucas Learning' on it. If we market it as an educational product, they don't want to buy it."

He explained how the company has been feeling a growing pressure to follow in the footsteps of established successes such as the *JumpStart* line that are cheaper to make and still sell well. I interviewed Blossom's colleague Michael Wyman while he was in the midst of production for *Pit Droids* in 1999. He mentioned that budgets have been shrinking in the kids' category and that the *Pit Droids* effort was probably "the last show and tell. I feel like I got really, really lucky to have the chance to spend the kind of money I'm spending and doing what I think is a really great product. Will it be profitable? I hope so, but I don't know." He worries that it will take major marketing muscle to bring attention to a new genre of the sort represented by *DroidWorks* and *Pit Droids*. Unlike a major toy company, Lucas Learning has worked to make quality the beacon for dissemination rather than a television ad campaign. As a result, although its products have gained critical acclaim and have been moderate sellers, they have not achieved the major commercial success of some of the Barbie and LEGO software products.

Construction, authoring, and simulation tools thus have a different marketing challenge, and their appeal is generally packaged as technical empowerment, the ability to translate authorial agency into a technologized form. "Create your own Star Wars world with fantastic 3D creatures," suggests the *Gungan Frontier* box. Such a position is easier to take for titles that more clearly fit the authoring genre (such as word processors or drawing tools) than for games such as *Gungan Frontier* that are presenting a new and unfamiliar concept. "Create anything imaginable in real-time

3D!" proclaims an ad for Disney's *Magic Artist 3D*, featuring a virtual Mickey stretching the surface of a shiny, three-dimensional, textured Mickey logo. The claim and the appeal focus on an ability to mobilize as substantial a technical apparatus as possible in the form of full three-dimensional, animated graphics. Like *SimCity 2000*, titles in this genre are often not age specific and cross over between adult and kids markets. "The power to change history is in your hands," announces the ad copy for the game *Call to Power*, packaged as being within the adult entertainment genre, but advertised in *Family PC*. Like *SimCity 2000*, this game is a strategic simulation that enables a user to author a virtual world. "*Call to Power* lets you create a world of your very own." Ads for *SimCity 3000* that ran in *PC Gamer* in the late 1990s similarly stress that radically different game outcomes can result from the player's actions (figure 4.1). "Mr. Rogers or Mr. Hussein?" queries the ad copy over contrasting images of a peaceful town or a blown-out office building. "It's a beautiful day in the neighborhood when you've got the power to rule over *SimCity 3000*." Such titles package their appeal with claims that they allow personal identification, customization, and authoring rather than with claims that they will transmit specific bits of knowledge or spectacular pleasures.

Software Case: *SimCity 2000*

SimCity and the simulation games that it ushered in are by far the most successful set of titles in the child-friendly constructionist genre. The most current game in this line, *The Sims*, is the best-selling game software in history. As a type of software that integrates popular culture and authoring, and as a title in wide use at the 5thD clubs, *SimCity* provides a focus for my discussion of the construction genre. Maxis, the company that produces the Sim line, started as a venture in 1987 between Will Wright and Jeff Braun to market *SimCity*, a game that Wright had designed and programmed. The two men deliberately avoided the label *educational software* for *SimCity*, believing that "people have a low opinion of educational software" (Wright in Barol 1989, 64). The *SimCity 2000* sourcebook bills the game as "entertainment/educational software" (Dargahi and Bremer 1995, 4). The game crosses the boundary between the explicitly educational children's market and the entertainment market, competing in the category of strategy and simulations games. The *SimCity* games are rare

Figure 4.1
Two advertisements for *SimCity 3000*. © 1998 Electronic Arts Inc. SimCity 3000, SimCity, Maxis, and the Maxis logo are trademarks or registered trademarks of Electronic Arts Inc. in the United States and/or other countries. All rights reserved.

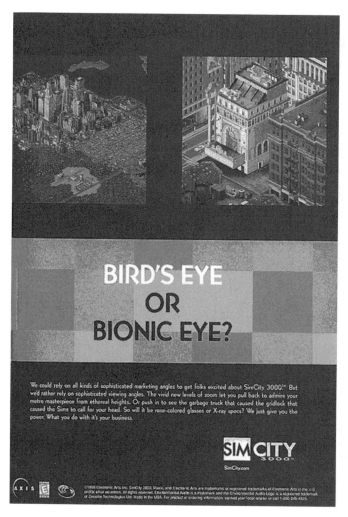

Figure 4.1
(Continued)

crossover hits, have won rave reviews from educators (Eisler 1991; Jacobson 1992; Paul 1991; Peirce 1994; Tanner 1993), and have attracted a wide following in the computer-gaming community as a whole. In 1997, Electronic Arts, the primary producer of action and sports computer games, bought Maxis and moved away from efforts in the children's and educational market. Until that time, however, Maxis had a division that specifically worked on children's titles and on educational framing of other Maxis titles. Maxis recognized the legitimizing aspect of the educational market, even while it catered to the economic strength of the entertainment market.

Maxis released *SimCity 2000* in November 1993, following a one-million-dollar development investment (Darlin 1994, 300). Unlike the original *SimCity, SimCity 2000* was the work of a large production team and integrated the suggestions of hundreds of fans who wrote in asking for new features (Dargahi and Bremer 1995, 337). *SimCity* had broken new ground by presenting an innovative model for a computer game based on world-building simulation, but *SimCity 2000* pushed the envelope on complexity and multimedia. It capitalized on the capabilities of new PC platforms, incorporating three-dimensional graphics and animation, advanced music and sound effects, new public-transit systems, a water system, hospitals, schools, and a complex new economic system (Dargahi and Bremer 1995, 396). Both *SimCity* and *SimCity 2000* were hit products in the competitive computer-game field, earning Maxis millions in revenue (Darlin 1994). The games have spawned an active subculture, with Usenet newsgroups, competitions, numerous publications, and even networked versions initiated by a loyal user community. *SimCity* is currently in its fifth incarnation with *SimCity 4.*

SimCity 2000, the focus of this chapter, is both authoring tool and interactive game. It provides a responsive virtual environment equipped with tools for users to build and administer a virtual city. The primary interface window is a grid that can be rotated or zoomed in and out, and that includes toolbars bristling with buttons and pop-up windows. Starting with a blank landscape dotted with trees, water, and hills, the player chooses different tools from a toolbar running alongside the screen, building, bulldozing, and zoning. In addition, informational windows report on population, citizens' education levels, pollution, industrial growth, and city budget, among many other factors. The game's basic progression involves building roads, zoning districts, and providing city services such as power, water, schools, parks, and libraries. In addition, the player must make decisions about budgeting: taxes, city ordinances, and allocation of funds. If the city is zoned and administered properly, "Sims" (simulated people) will populate the grid, creating their own buildings and voicing their opinions through the city newspaper.

The user plays the role of city mayor and receives rewards for good governance and population growth, such as a mayor's house, a statue, or a spontaneous parade. The system promotes a model of expansion and growth through rewards for achievement of population levels, but parallel and subgoal structures exist in the game, including ecological and economic balance, community relations, and aesthetics. Because *SimCity 2000* foregrounds user authoring, it is less a game driven by a specific goal than a structured space of possibility for the user to explore. Disasters—including fires, floods, and space alien invasions—can be turned on and off to subvert linear-growth scenarios. The algorithms underlying *SimCity 2000* rely on cellular automata techniques,[2] creating an impression of lively growth, interactivity, and change—a sense of the city as a living entity. Construction sites change to small buildings, which are in turn torn down to make space for a large shopping mall or a stunning skyscraper. As the city's population and transportation network grow, antlike cars start flowing frenetically across highways and roadways, and planes and traffic helicopters fly across the cityscape, occasionally crashing into a tall building and maybe even starting a fire. The disaster function also creates periodic floods, fires, earthquakes, and other calamities that spread unexpectedly through the city. The focus of the game is to create a viable and aesthetically pleasing simulation. Goal orientations are afforded by the rewards functions that give players not only kudos for achieving population levels in their city, but also the persistent possibility that their city can fail by going bankrupt or becoming unlivable for the Sim citizens.

The interactional dynamic is much like engaging with a drawing program. The player chooses a tool, such as a zoning tool, a bulldozer, or a tool to build a specific public building (e.g., stadium, park, or power plant) and "draws" that element onto the grid. Auxiliary windows such as the budget window and city ordinance window allow the player to adjust elements of the city, such as tax rate or monies allocated to city services, which are not represented visually on the primary window grid. In contrast to a drawing program, however, the software will "interpret" player input and perform operations on that input in the form of simulations, such as populating a newly zoned area or having Sim citizens flock to a new stadium. By integrating an urban-planning scenario and an authoring tool, the game produces the effect of a godlike bureaucrat, able to produce immediate outcomes in a city by virtue of simply laying out a city plan.

In contrast to many games adapted from noncomputational and narrative media (e.g., animated storybooks, adventure games), Sim games tend to be functionally complex and allow players to engage and tinker with the technical substratum. They allow players to exercise what Ian Bogost calls "procedural literacy," or "the ability to reconfigure concepts and rules to understand and process, not just on the computer, but generally" (2007, 241). He suggests that *SimCity* not only exposes players to certain content domains, but also enables them to engage with underlying procedures and processes that are specific to the domain, whether taxation, ecology, or urban growth. The special effects and graphical representations at the interface are generally tied into substantive functionality that is relevant for game outcomes. For example, the flow of cars across a freeway is an indicator of traffic patterns, which are tied to other game factors such as the Sim citizens' level of satisfaction, pollution levels, and population. In contrast, in games that rely on storyline and characters for their narrative cohesion, graphical representations may be meaningful for the player, but not for the underlying machine logic. Although providing a robust fantasy environment, Sim games are characterized by the dominance of technical and procedural logic over narrative logic.

The cellular automata algorithms are not directly available to the player, but numerous supplementary materials for the game enable the player to delve below the interface elements and engage with the nuts and bolts of how the city runs and grows. The game manual and lengthy "sourcebooks" document the algorithmic relations that drive the simulation. For example, the Sim citizens are programmed to walk only a certain number of game squares to get to public transportation. Otherwise, they will take their cars. Algorithmic relations of this sort, described in supporting materials, determine many of the contingencies for the city: the life span of power plants, the relations between transport systems and pollution, the consequences of city ordinances, and so on. Technically savvy players can purchase the *SimCity 2000 Urban Renewal Kit* that enables them to design their own buildings and tinker with the algorithms that drive the city. *SimCity 2000* includes some of the more complicated and dense layerings between narrative, interface, and functionality that were possible in what was state of the art at the time it was released.

SimCity 2000 in the 5thD

SimCity 2000 has had some persistence at the 5thD, and our data covers both the 1994–1995 and the 1995–1996 school years. It was introduced into the 5thD in February 1995 with fanfare as a gift at the Wizard's birthday party, the major celebration for the club's year. The game was a special request from Ian, the boy who appeared in the previous chapter. Ian was a great fan of the original *SimCity,* a game well established as a popular game at the site at the time of *SimCity 2000*'s introduction. *SimCity 2000* was installed on the machine for exclusive use by the Young Wizard's Assistants, but regular 5thD citizens were soon petitioning the Wizard to be able to play the game and were using their "free passes" to gain access to it. Because the game was so popular, kids began a practice of meting out game play in thirty- or fifteen-minute turns so that no one kid was able to play for the entire duration of the kids' time at the site. The game also became a motivation for kids to achieve Young Wizard's Assistant status, and field notes document how at least a few kids were working diligently through the site maze for the purpose of gaining access to *SimCity 2000*. The *SimCity 2000* machine was almost always surrounded with a small group of two to five kids, usually the older boys (ages ten to twelve), some of whom were onlookers (both kids and undergraduates) and others who were waiting their turn. Most observers, especially if they had some knowledge of the game, were not shy about throwing out suggestions to the player and critiquing the player's city.

Many undergraduates comment in their field notes on the "spectacular" or "incredible" graphics and sounds, and describe the game as "ever popular," "attractive," and "exciting." Kids were generally enthusiastic about the game, as evidenced by their maneuverings to get a chance to play as well as by their descriptions of the game. Undergraduate notes document how *SimCity 2000* was a constant magnet for the kids. In response to one undergraduate's asking if they liked the game, one boy responded: "Are you kidding? We're Sim kind of people. Sim City, Sim Kid, Sim Ant." "I love it, definitely," another kid replied. During the course of game play, they often commented on a "cool" graphic or sound and would show off innovations in their cities to other kids and undergraduates. One boy, insisting that he wanted to play a "noneducational game," suggested *SimCity 2000*.

Overall, kids who played *SimCity 2000* in the 5thD had a strong sense of ownership, and identified it as a game that catered to their, as opposed to adult, agendas. Undergraduates typically were attracted to the game as visually appealing and challenging, but unless they were experienced with the game, they were generally put into the observer role. They often described their confusion. As one undergraduate stated after her first day with *SimCity 2000*, "Since I do not have the slightest idea as to how *SimCity 2000* is supposed to be played, I was unable to provide . . . any help." Even after a second day observing the game, she wrote similarly: "Since I don't know how to play the game and have no suggestions or advice to offer I usually just sit back and watch the children play." This undergraduate's experience was not unique; when there were a group of kids playing the game, the undergraduates' role was almost always one of peripheral observer rather then helper or coparticipant. Because the undergraduate cohort was renewed each quarter, kids achieved a degree of expertise by the end of the school year that the undergraduates couldn't and were almost always in the position of teaching each other as well as the undergraduates.

When an undergraduate with some experience in playing the game did turn up at the 5thD, he was put in the position of having to demonstrate his expertise before kids would listen to his suggestions. One undergraduate who was playing with Ian suggested using nuclear rather than coal power to reduce pollution, and Ian challenged her, "How do you know that?" When she explained that she saw it in the user manual, he asked for confirmation: "Are you sure?" It was only after her confirmation that he took her suggestion. Undergraduates sometimes tried to bring in knowledge about the world at large to inform game play. "I suggested to Jimmy to put in some schools as there were none in the city. I told him, 'People will stop coming to live here if there aren't any schools.' . . . He looked doubtful."

The kids and adults' orientation diverged in other ways as well. The kids were more versatile in their play, engaging in goal-directed play as well as in more open-ended play with the game's aesthetic dimensions. By contrast, the undergraduates generally tried to get the kids to play toward city growth and management. One undergraduate mentions her interactions with one boy who was using the game as an exercise in "creative design": "I suggested he play the game correctly to see if he could build an efficient city that worked, but he told me he was having more fun just designing it." Playing with disasters, described in the previous chapter, is also part of this more exploratory mode of play. Another undergraduate writes about how one boy "was frustrated and only wanted to blow up a city. . . . We both protested and finally [another boy] came to control the mouse and we were able to get him to begin to be rational in his decisions as we had been doing." When a different undergraduate suggested that the boys reflect on how to make their city work better, she was ignored. "I think they were more interested in the plan of the city rather than the success of the city." The boys she was working with focused instead on the placement of the mayor's house. Undergraduates were also surprised at kids' lack of investment in their cities: "When I looked back at the game, Kevin had DELETED the whole thing ON PURPOSE. He had just spent like an hour building this city." Another undergraduate notes with dismay how the boy she was working with "did not want to save his city, and shut off the game." One onlooker suggested, "I would have destroyed my city with a natural disaster like an earthquake or a hurricane."

Given the game's complexity, most kids who played during the 1994–1995 year, when the game was introduced, only scratched the surface of its capabilities. Without an experienced coach, they often failed to uncover the game's basic functionality. Without recourse to the game manual or an experienced player, it is not obvious how one begins a city and gets it populated; the player must have roads, three kinds of city zoning, a power plant, and power lines just to begin to populate a city. When working with an inexperienced undergraduate, kids usually began exploring the complex toolbar that allows players to build city infrastructure and public buildings such as police stations, stadiums, and hospitals, and would often build these parts of the city without knowledge of the minimal conditions to attract a city population. The player would thus soon run out of money and have to start a new city. Kids and undergraduates also spent much time puzzling over the actual interface, looking for the button for the power plant, trying to figure out how to build roads up a hill, wondering how to connect a set of train tracks, or trying to build on on-ramp to an otherwise nonfunctional freeway. Given the constant circulation of kids around the *SimCity 2000* machine at the 5thD at this time, however, the more experienced kids would often chime in and explain that a road needed to be built or that a power line needed to be drawn to a certain area, and most kids, by the end of their session, were able to attract at least some Sim citizens.

After some citizens are finally attracted to the city, the next challenge is budgeting. The game is designed so that it is easy to run out of money in the first hour of game play, so avoiding this outcome requires careful planning and some understanding of how to manage the city budget. Early on, the practice in the 5thD was for kids to take out bonds as soon as money ran out and to prolong play time before bankruptcy. As the bond payments increased, however, they would inevitably run out of credit and money, so they would just start a new city, often after destroying their bankrupt city with a series of disasters. It was only after an undergraduate experienced in the game appeared at the club a few months after the game was introduced that kids were able to move beyond the initial troubles with budget management. The strategies for moving toward a positive balance sheet are complex and include reducing funding for certain city services, adding or reducing certain city ordinances, taxing different industries at different rates, pursuing controlled city growth, and constantly monitoring property tax rates. The field notes and videotape document the gradual increase in expertise on the game at the club with the advent of the undergraduate expert and the knowledge about the game that the kids began to bring in from home. It is unclear whether kids and undergraduates, without the help of experts of this sort, would have succeeded in overcoming the basic problems of budget stability and population growth.

Despite the persistent difficulties in building and maintaining a stable city, almost all kids continued to play and enjoy the game. In addition to the game's goal structure that stresses city growth, the kids often oriented toward its aesthetic and design dimensions, spending much time building elaborate waterfalls, twisted subway systems, lakes with multiple marinas, enormous airports or seaports, islands shaped like letters, and a private mayor's retreat. They also were strongly attracted to the countless disasters that are only incidentally related to the game's more functional goals. In the fall of 1995, when the 5thD started up for a new school term, one of the kids experienced with the game arrived with a new bit of knowledge that furthered this tendency toward play with the aesthetic and "destructive" dimensions of the game. One of his older friends had divulged to him the secret code that gave the player unlimited funds and access to all of the special buildings that are generally awarded only after a player's city has achieved certain population levels. In other words, the code enables the player to circumvent the game's structures that encourage attention to fiscal issues and population growth. After this knowledge became disseminated at the site, kids new to the game would struggle for a while with budget and population issues, but sooner or later the secret was divulged to them, and they thereafter tended to use the game more as an elaborate authoring tool.

Power Users and Hard Mastery

SimCity 2000 is a game with multiple technical layers that the player can manipulate. In contrast to educational software that seeks to convey a set knowledge domain, often in narrative form, authoring and simulation games require the player to master a set of technical tools to design customized narratives and meanings. A certain amount of technical knowledge is a requirement of play. The "learning narratives" about software such as *SimCity* and *SimTower* generally explicate procedural knowledge and technical functionality: how to place a marina or a bridge in relation to land and water formations, how to schedule elevator routes, how to manipulate tools to make the subway system connect, what the relationship is between zoning and population density. There is often a sense of accomplishment associated with learning how to use new tools and discovering different technical functions and relations, which differs from the sense of achievement felt in "beating" a game or completing a set of problems.

A few weeks after the game was introduced, an undergraduate, Bruce, a power user of the game, appeared at the 5thD and began to reorganize kids' play toward more sophisticated technical mastery and new forms of procedural literacy. In his field notes, he describes how the kids had been taking out bonds that the city would never be able to pay back, dooming the city to eventual fiscal failure. He sat down with the Young Wizard's Assistant who had been most involved with the game and gave him detailed tips on how to balance the city budget. He showed him the window that allocates specific amounts to different city programs and how to fine-tune these settings to maximize revenue. "One of the main strategies is to cut all funding to the fire department and turn off the 'disasters' mode. When this mode is turned off, there can be no fires, so fire department funding is not necessary." Working on a subsequent day with another boy, he wrote that the boy "was a very fast learner, and genuinely interested in the more complex aspects of the game. It's fine to put the game on 'Autobudget' and watch it slowly grow, but he, like me, found it more fun to play around with the city's budget and taxes to increase income." Bruce worried, however, that he might have initiated a technical discourse that would be over the head of some 5thD participants: "I can just hear him approaching an eight-year-old just learning to play the game and saying 'No! You have to cut the spending to the fire department, cut spending to un-needed transportation system, increase taxes on textiles and steel, decrease taxes on automotive, implement sales tax and a smoking ordinance, and don't forget nuclear-free zones! Don't you know how to play this game?'"

One example of Bruce's teaching of *SimCity 2000* functionality illustrates the sorts of interactions that I am characterizing as oriented toward mastery of the game's technical logic. As shown in the videotaped session, Bruce is playing with Dean, and Ian is observing them. Bruce points out some of the more complicated features of the game to Dean and triggers some talk around deciphering a particular graph.

D = Dean
I = Ian
UG = Undergraduate (Bruce)

1   UG: You see this little graph? (Points.) You know what that means?
2   D: (Opens graph window.)
3   I: Whoa! Go back, go back (responding to the window's snapping shut).
4   D: (Opens window and drags to keep window open.) That's my S. That's my something. That's my industrial. That's my residential, and that's my commercial. (Moves cursor down to point to different lines in the graph, labeled "S," "I," "R," and "C," as he refers to them.)
5   I: Commercial.
6   UG: Right. This other one right here. (Points to another item on toolbar.)
7   I: (Jumps up and points to closing window.)
8   D: (Closes graph window.)
9   I: The seaport! That's seaport! That's seaport!

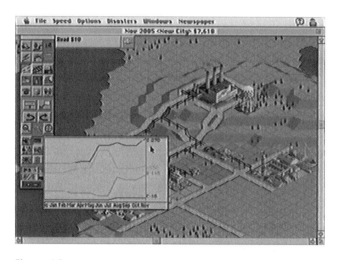

Figure 4.2
Graph of zones in *SimCity 2000*. Screen shot reproduced with permission from Electronic Arts Inc. © 1993–1994 Electronic Arts Inc. SimCity 2000 and SimCity are trademarks or registered trademarks of Electronic Arts Inc. in the United States and/or other countries. All rights reserved.

In response to Bruce's question, Dean makes links between the lines on a normally hidden graph window and the different zones in his city—industrial, residential, and commercial (figure 4.2). He skips over the line labeled "S"—"That's my something" (line 4)—a move that Bruce also glosses over. Ian, however, notices the omission, and as Dean closes the graph window, Ian jumps up and enthusiastically points out that "S" must refer to "seaport," a special kind of zone (line 9). The excitement stems from his uncovering a new symbolic relationship between two game elements. His insight is not, however, acknowledged by the other two, who move on to the next graph that shows population levels. Even though the others do not take up Ian's contribution, the three participants have managed to cement a common interactional space focused on deciphering the intratextual referents within the game, the meaningful space of identifications looping tightly within the game's technical terms. The narrative logic of learning "about seaports" is secondary to the sense of revelation in uncovering a representation tool. In this scene, Bruce, Ian, and Dean seek to uncover the particular instrumental logic of the game as designed by the Maxis production team. Like film critics that might analyze a camera angle and lighting, the three forge links with the production apparatus, making the game meaningful as a complexly embodied technical achievement. This kind of interaction, where kids and undergraduates work together in figuring out how the tools and interface work are frequent in the transcripts for games like *SimCity* and *SimTower*.

As kids gain more technical expertise, they can push the technical functions even further. In another instance of play recorded some months later, Ian is playing with a different undergraduate who is not as familiar with the game. In field notes, the undergraduate describes the kind of experimentation Ian was engaging with, testing the parameters of the simulation:

Ian wanted to try out his *SimCity* experiments so he closed and saved my city. He wanted to see how well a certain type of arcology would survive in a series of earthquakes. He created a number of these mini-cities all over the terrain then clicked on the Disasters menu button and proceeded to subject them to 30–40 earthquakes. I had a feeling of being witness to a scientific experiment: I also wanted to see what would happen. I told Ian that arcologies were a little like the city-states of ancient Greece and I explained that the "polis" served as an early form of democracy where men, not women at the time, could vote and run the city. The arcologies we were testing did not all crumble at once. After 10 earthquakes, 2 collapsed. After 30, most of them fell.

Within a few months after the game's introduction at the 5thD, a strong peer culture had arisen around the game, dominated by the older boys, ages ten to twelve, who frequented the site. The regular girl participants at the site also enjoyed the game, but they rarely participated in the group interactions around the game. One excerpt from a field note written in the second school year after the game's introduction describes the nature of the interaction between the boys:

There were about five boys crowded around one terminal when I first joined the game. Another [undergraduate] was following along with the boys also. George was seated in front of the terminal controlling the keyboard, but all of the boys were shouting out suggestions and reacting to the things George was making happen on the screen. . . . The boys spoke loudly to each other, but they weren't fighting. They were more or less challenging each other's expertise on the game. If one of the boys told me something that the others disagreed with, a debate would immediately take place until it was resolved. Each of the boys was extremely informative when I had a question. They would tell me everything that they knew, even if it was way past what I had originally inquired about.

Although the undergraduates often commented on kids' willingness to share their knowledge at length, they also at times felt left out by their own lack of understanding. In her book *Cheating: Gaining Advantage in*

Videogames, Mia Consolvo (2007) discusses how the circulation of insider knowledge about games constitutes a form of gaming capital that confers status and expertise in the subculture. The peer culture in the 5thD that centered on a more "geeked-out" form of engagement (Ito, Baumer, Bittanti, et al. 2009) with *SimCity 2000* quickly built up forms of esoteric knowledge that excluded outsiders. As one undergraduate notes, "Another child by the name of Mick stood beside me and talked a lot about the game, most of which I couldn't follow because they [his explanations] were so esoteric." Although I don't have any direct evidence that girls were turned off by this "in group" of *SimCity 2000* boy players, girls' lack of participation in these group discussions is evidence of a gender bias, wherein boys dominated these technical debates. The instances of girls' play we have on record are all in situations where one girl is playing with an undergraduate or with one other girl friend. One girl at the site did take a keen interest in the game and started working diligently toward becoming a Young Wizard's Assistant so she could gain access to the game. In contrast to Ian's intense and ongoing engagement, however, she grew bored with the game soon after she becomes an assistant. An undergraduate comments, "She blew her budget on things not really needed. She called her town exclusively La Jolla in her first three attempts but as she started giving up on the game she was calling her town Dumb Town. She was giving up without really trying." There are no records of her playing with the game after this instance of play.

Creating My Own World: Identification and Self-Actualization

Competitive and spectacular pleasures are present in simulation kits, but differ from the pleasures involved in creating and authoring a unique virtual world. Authoring systems such as *SimCity 2000* allow players to create their own spectacles, settings, characters, and interactive possibilities—to construct user subjectivity as a world builder rather than a world explorer. Game play and mastery involve uncovering the technical functionality of the building tools and then executing a personalized vision of a city while balancing between different factors such as cash flow, population density, and aesthetics. Just as children learn the relation between materials and physics when they manipulate building blocks, *SimCity* mastery is about being able to manipulate and combine computational building blocks into a unique structure. In contrast to their absence in discussions of power use, girls are often active participants in these creative activities (Ito and Bittanti 2009).

The results of these construction endeavors are often highly invested with personal reference, style, and meaning—a sense of creative *accomplishment* that differs from a sense of *achievement* in more clearly competitive scenarios. Kids at the 5thD continuously debated what features and buildings were cool and where they should be placed, and then showed off their creations to one another. "My friend had one of those big shark-looking things up on a big hill that he had made!" "Look! Want to see my huge mountain? I made this huge mountain by hand, and I covered it in water by hand. I did this all by hand." When playing in a group setting, negotiating authorship was often difficult because of the unique and personalized nature of the choices being made. In the record of one interaction, a female undergraduate and two boys debate whether to put in a zoo or a park in a location by the mayor's house. Mark, the youngest of the group, is in control of the mouse, helping build a city for the undergraduate.

UG = Undergraduate
I = Ian
M = Mark

1   UG: Why don't you put in a park?
2   I: Put in a zoo.
3   M: I want to put in a stadium. [Picks stadium.]
4   I: Dude, what did you want?
5   UG: It doesn't—let's put in a zoo.
6   I: She [the undergraduate] wants a zoo.
7   M: I want a stadium.
8   I: She wants a zoo. It's her space.
9   M: I want to do a stadium.
10   I: She wants a zoo.
11   M: You said I could do something.
12   UG: Here. You can put in a stadium. We'll put in a zoo later, OK?

In this sequence, the undergraduate eventually compromises. "You can put in a stadium. We'll put in a zoo later, OK?" (line 12). Her decision stands because they had started the city as "her" city. Cities include others' input but are experienced as personal creations.

One way that *SimCity 2000* accomplishes a sense of identification is by suturing the player into identification as the mayor of the city. Personnel in various city offices advise "you" on the state of transportation, education, and other city services. After achieving a certain population level, you are awarded a "mayor's house" that you can place at will. This moment appears frequently on tapes of children's play as a key moment of identification. They generally spent a great deal of time working to place the house in a nice location—on top of a hill, overlooking a lake, distanced from the bustle of the city—and sometimes added a private subway or park for it. The player with ownership of the city without exception referred to the house as "my house," and the appearance of the house generally initiated a sequence of imaginative projection, where interlocutors talked about what it would be like to live in a particular location. Even older players enjoyed this kind of identification with the game. "Girl, you want to see it?" one of our teenage players asked a new friend at the club. "Watch this . . . watch. That's my house, girl." "Oh!" proclaimed the friend, impressed. "I want to build. I'll make mine."

One instance of a boy (Jimmy) and an undergraduate (Holly) discussing the mayor's house is representative of these interactional moments. They have been working on this city for about twenty minutes, with Jimmy in control of the mouse the entire time. Their city has just reached a population of one thousand, which triggers the "rewards" button to highlight, indicating that the player can build the mayor's house. As Jimmy moves his pointer up to select a button on the toolbar, he notices the rewards button, and the following exchange ensues.

J = Jimmy
UG = Undergraduate (Holly)
square brackets signify overlapping talk

1   J: I can make my house, the mayor's house. (Clicks on rewards icon.) Where do I want to make my house?

2   UG: (Laughs.) You want it overlooking everything? (Laughs.) Aaa . . . Do you want to have it overlooking the lake or something?

3   J: (Dismisses year-end dialog box.)

4   UG: Yeah, you can have like those be the really nice houses or something. Like up in the hills?

5   J: Up here? (Moves cursor to flat area on ridge.)

6   UG: Uuuuhhh, maybe over here (pointing) because this you'd be just overlooking over the power plant. That wouldn't be very nice. Maybe over by the lake or something?

7   J: How about right here? (Positions cursor over flat ground by the lake.)

8   UG: Sure, yeah, like right on the lake?

9   J: (Builds house on lake opposite city.) Yeah, I need power, *obviously* because it's my house.

10   UG: (Laughs.)

11   J: (Builds power lines to mayor's house.)

12   UG: You need water I would think.

13   J: Hate water! (Selects water pipe tool.)

14   UG: (Reads from status bar.) Water shortage or something. Why would there be a water shortage? Something like a drought or something? You can't really . . . mmm . . .

15   J: (Builds pipes to mayor's house.) There! (Water keeps running out.) Better give me water! (The water doesn't flow to house.) I don't care, though! (Dismisses budget window.) It's so hard . . .

16   UG: What is?

17   J: Using this. . . . (Selects hospital tool.) Hospital? Should I put in a hospital?

18   UG: Another one?

19   J: Do I already have one?

20   UG: Uh-huh. You have one by the college. You have another on . . . Do you have a free clinic or something for the people [who can't afford it?]

21   J: [How 'bout auuummm!] A prison.

22   UG: You have a police station already, right?

23   J: Where should I have it, right here?

24   UG: That's over by the hospital. You probably want it . . .how about over by the sewage, like the industrial area. So it's not, 'cause you don't, I mean, like, no one would want to live in that area.

25   J: (Tries to place prison on terraced ground.) I can't put it on the, like, it has to be on flat ground. How about right here? (Positions cursor by lake.)

26   UG: Don't put it right by your house! You don't want to live by the prison, do you? (Laughs.)

27   J: (Laughs.) Right here? (Positions cursor on opposite side of lake.)

29   UG: Sure. That's right by the police station. You might as well.

30   J: More convenient.

In this segment of activity, Jimmy begins by taking up an identification between himself and the mayor, as proposed by the game—"I can make my house. My mayor's house"—and then invites Holly into the decision of where to place the house: "Where do I want to make my house?" (line 1). Holly's talk then draws in a series of connections from her knowledge about the world: what constitutes desirable real estate, a good view, and signifiers of power and wealth (lines 2, 4, 6). Jimmy takes up her first suggestion, to put the house overlooking a lake, by trying to place it on a ridge

Figure 4.3
Screen shot of Jimmy's mayor's house in *SimCity 2000*. Reproduced with permission from Electronic Arts Inc. © 1993–1994 Electronic Arts Inc. SimCity 2000 and SimCity are trademarks or registered trademarks of Electronic Arts Inc. in the United States and/or other countries. All rights reserved.

overlooking an industrial area. She then offers an alternate suggestion, to place the house by the uninhabited region by the lake. Jimmy, with her agreement, places "his" house by the lake opposite the city (figure 4.3). They go on to consider the placement of a prison, and Holly advises against placing it next to the mayor's house (lines 23–25). She is not basing her suggestions on any knowledge of the underlying algorithms of the game that calculate real estate value or the "not in my back yard" (NIMBY) phenomenon, but rather on what makes the game sensible for Jimmy in terms of her sociocultural knowledge. In fact, the placement of the mayor's house is inconsequential in terms of game outcomes, but functions as an effective hook for locating a subject position for the player within the game's mis-en-scène and, conversely, for locating the game within a system of social distinctions. This kind of talk, where Holly refers to social contexts at large to make sense of the game, is fairly typical of their interaction. At other times, for example, Holly initiates discussions around what kinds of transit systems might be desirable or how good school districts attract families. Although the tone of the talk is decidedly playful and peppered with laughter—it's just a game, after all—Jimmy, Holly, and *SimCity 2000* have succeeded in organizing themselves around a series of identifications that link Jimmy, the mayor, "nice" parts of town, and a good view.

In another instance, two girls, Allie and Jean, are playing the game with an undergraduate. The undergraduate's field note describes how they engage in a series of imaginative projections where they build a mansion for themselves and then a home for their parents far from their home. They continue to build homes for others in their lives.

There is a young man named Seth who Allie doesn't like, and she built him a house on the far corner of the city, and proceeded to surround his home with dense forest, far from civilization, roads, and water. Because neither girl likes Jean's four-year-old sister, Allie built her [the sister's] house in a deserted valley without water or electricity, to die in isolation. I said that it was mean, so Allie placed a teeny, tiny lake far away from the sister's home.

Children who engaged with *SimCity 2000* or *SimTower* over the course of multiple days developed even stronger attachments to their creations. One of the girls in our teen club built her tower over a period of weeks. On her first day of play with *SimTower,* in a dull moment while she was waiting for income so that she could build more of her tower, I had shown her some of the game's peripheral features. If the player clicks on a particular person in the tower, she can assign a name to that person. People named in this way appear in blue instead of the generic black and can be "found" through the "find person" command, where an arrow on the screen will point to where they are in the building, whether they are in their office or having lunch or waiting to get on an elevator. On a subsequent day, as captured on video, Brandy is displaying her tower to a visiting teenager, Kathy, and showing her how she has named people in her tower after her mother and her best friend, Tamika (figure 4.4).

B = Brandy
K = Kathy

1   B: See, now I'm out of money, so I have to wait until all my money goes up because I just wasted all my money on another elevator, and I have people that I named, watch. See who's on this floor? Nobody I know. See the people I named, they're blue. They're . . . look it, watch. (Clicks on the coffee shop, then on one of the blue people in the coffee shop.)

2   B: See all the blue people? And I have to click on them, and that's [Stella], that's my mom. She's at work, and she works on the office floor two, so that'd

Figure 4.4
Screen shot from *SimTower*. Reproduced with permission from Electronic Arts Inc.
© 1995 Electronic Arts Inc. SimTower is a trademark or registered trademark of
Electronic Arts Inc. in the United States and/or other countries. All rights reserved.

be that floor. Look, let's find [Tamika]. Let's see if [Tamika] is at work; if she's
not, she's getting fired. (Goes to the "find person" command and finds Tamika.
The window listing named people pops up.)

3   B: Let's see. These are all the people that I named. There's [Tamika's] name
         right there. I have more people, though.

4   K: How many did you name?

5   B: All of them.

6   K: Cool.

7   B: See, [Tamika] is there. She's got to be there. I think she works on the third
         floor. Yeah, oh, she's eating. So that's [Tamika] right there.

8   K: She's eating?

9   B: (Clicks on Tamika.) Yeah, she's, like, getting her lunch.

10   K: Cool, [Tamika], you're eating.

11   B: You're getting something, you're eating, uhm, what are you eating?
         [Tamika] is eating, you're eating Japanese soba, soba, soba.

12   K: How do you know what she's eating? Oh my gosh.

13   B: 'Cause that's what the restaurant is. Let's see who this is, [Tamika]? (Tries
         to click on a different person in the tower.) Catch them. Oh man.

In this sequence of activity, a peripheral element of the game that is
inconsequential in terms of building a smoothly operating tower, becomes
a social resource. The ability to name people in the tower enables Brandy
to translate game elements into relationships that are part of her everyday

life. This process not only links her particular relational universe (best friend and family) (line 2) into the space of the virtual world, but links these familiar aspects of life with more unfamiliar ones. The game becomes a vehicle to imagine what it might be like to be creator and administrator of a skyscraper and for a teenage friend to work in an office, get fired, or eat Japanese noodles (soba) for lunch (line 10). It is a source of pleasure not just in engaging with the visual and interactional effects of the game, but in forming a personal identification with and a sense of ownership over a unique creation.

Engineered Subversion

The previous chapter introduced a sequence of play where Ian, a boy heavily invested in *SimCity 2000,* is playing the game with some other kids and is asked by the site director to teach a new undergraduate how to play. The sequence of play illustrates not only the effects of children's peer networks in engaging with special effects and consumer culture, but also some of the shifts in consumption-production relations when kids are handed a flexible set of computational authoring tools.

Ian, the eight-year-old who forms the basis of this case study of *SimCity 2000* use, was a veteran of the 5thD at the time and a Young Wizard's Assistant, which meant that he had completed all the games in the maze and had earned the right to play the high-end games at the site. It also meant that he was responsible for teaching others. At school, Ian was flagged as a problem child and diagnosed with attention deficit hyperactive disorder, and at home he was subject to a behavior-modification schedule. At the club, however, he was known as a game expert who could control the attention of both other kids and adults. Undergraduate field notes frequently describe him as exceptionally bright: "I was very impressed with his knowledge; he seemed wise beyond his years." Despite his identity in the club as a game expert, undergraduate field notes also document how Ian didn't think he was a smart kid. When an undergraduate complimented him as "smart" in relation to his game expertise, he reacted with surprise, saying that he didn't think he was smart because he had been held back in school. Like Roger, a central figure in chapter 2, Ian was an uncommon kid who appeared at the center of certain social tensions in the 5thD. He challenged the school and club's adult educational agendas by mobilizing technical resources in unexpected ways that didn't conform to those agendas.

Ian was particularly fascinated with *SimCity* and *SimCity 2000* and was known for his intense engagement with these games, usually at the expense of learning other games or working with others at the club. His letters to the 5thD Wizard and the undergraduates' field notes reveal his constant negotiation, pleading, and wheeling and dealing in order to play *SimCity 2000.* Before becoming a Young Wizard's Assistant, Ian was largely excluded from playing *SimCity 2000* because it was a game reserved for the assistants or for kids with special permission from the Wizard. During this period, he barraged the Wizard with special requests and pleaded with the site coordinator to be allowed to borrow the software to play at home. Videotape of this period shows him lurking at machines while other kids play, tossing in suggestions, and futilely pleading to be able to play. After he became a Young Wizard's Assistant, largely from his motivation to have unrestricted play with *SimCity 2000,* the struggle was to disengage him from the game enough so he could teach others how to play. These kinds of struggles, where Ian emerged as a devoted and engaged member of the club while persistently pushing against club rules, was typical of his 5thD identity. His hackerlike expertise and obsessive engagement with the game was cause for both celebration by adults as well as their constant efforts at redirection. Even after Ian achieved Young Wizard's Assistant status and gained free access to the game, adults intervened in his play by instituting a "twice-a-week rule" so that he would also play with other games.

In the previous chapter, I described Ian's "disaster time" wherein he subjected his city to various disasters—floods, plane crashes, earthquakes, fires, and explosions. After he was finally displaced from his burned, flooded city at the adults' insistence, he started a new city for the under-graduate he was supposed to be helping learn the game, and they, along with another kid, worked on it for forty minutes. During this period, Ian returned to construction mode, building buildings, power plants, the mayor's house, a railroad, and a subway system. Significantly, however, he began this new city by typing in a secret code that gives the player unlim-ited funds and opens access to special rewards, such as space-age buildings, the mayor's house, and all of the high-tech power plants. By typing in the secret code, he could circumvent the game parameters that demand atten-tion to fiscal responsibility and gradual urban growth. The game was transformed from an urban-planning exercise to a palette for the free construction of any desired elements.

Ian used the backdoor code as a way of expanding the space for personal agency, and he worked toward building a network of co-conspirators who would reproduce this alternative mode of game play. In the following excerpt from the tapes of his play, he has just been working to build a subway system, but has difficulty and needs to keep bulldozing and reconstructing. This process leads to a discussion of how much money he has wasted, but how it doesn't matter because of his use of the secret code.

I = Ian
UG = Undergraduate

1   I: We wasted hundreds and hundreds of dollars. I don't believe it. We just wasted about five hundred thousand dollars trying to connect it [the subway], and it was already connected.

2   UG: Whoops. Oh well.

3   I: That was a big mistake.

4   UG: That's OK. We still have tons more money.

5   I: Yeah, tell me when you want some more. More!

6   UG: Are you going to show me how? You're not going to show me the secret? Why not?

7   I: Promise you won't tell anybody?

8   UG: I won't tell anybody.

9   I: OK. Porntipsguzzardo.

10   UG: What did you push, what did you press, redtips?

11   I: Porntipsguzzardo.

12   UG: Wait, I don't remember.

13   I: Then you keep pressing "guzzardo."

14   UG: Where'd you learn that?

15   I: Somebody taught it to me.

16   UG: So you go . . .

17   I: Every time you type that, it gives you another half-million dollars. (Continues to type *guzzardo*, which continues to add money as citizens cheer.)

18   UG: Oh, wow. I don't think I need any more. Wow, they're cheering up a storm on the screen. Uh, look at how much we have. I don't think we need any more.

19   I: That's not very much.

20   UG: Not very much? So it's "porntips," then how do you spell the last, *guzz*—

21   I: Guzzardo.

22   UG: Guzzardo.

23   I: Guzzardo, double Z.

24   UG: Double Z. Thanks. Now I won't be . . .

25   I: (Continues to type code.) Is it changing right now?
26   UG: Yeah, totally.
27   I: How much do we have?
28   UG: Here, we have enough, we're at twenty million.
29   I: That's not very much. I had twenty-eight million.
30   UG: Twenty-eight million?
31   I: (Continues to type.)
32   UG: You just hear them screaming and screaming. They're going to lose their
             voices, they're screaming so much.
33   I: Twenty-nine.
34   UG: That's, that's plenty. You want to just go up to thirty? That's good.
35   I: Good.
36   UG: We have so much money that we won't even know what to do with it.
37   I: I know.
38   UG: That's so much money.

In this sequence, Ian has started to run out of money from building his subway (line 1) and asks the undergraduate when she wants more money (line 5). The undergraduate had previously noticed him typing in a secret code to get free funds, so she takes this opportunity to ask him if he will show her (line 6). Although resisting a bit at first, he tells her to promise not to tell anyone (line 7), takes a quick look at the video camera, and shows her the code (lines 9, 11, 12). He then continues to type the code until they reach thirty million dollars, an enormous sum in the *SimCity 2000* economy (lines 34, 38).

Ian's (not so) secret transmission of illicit knowledge was not restricted to this one instance on tape. On another day, he was asked to visit a neighboring after-school club to teach kids how to play *SimCity 2000*. His first act upon arriving at the club was to teach the kids the secret code, thereby subverting the possibility that they might engage with the game as an educational urban-planning simulation. He went on to show them the coolest buildings, and they experimented together on pushing their city to extremes—painting their initials in land formations, seeing if their city would survive various disasters, building a prison fortress reminiscent of Alcatraz, and building an enormous airport. He showed them various aspects of game functionality, how to zoom in and out and rotate the grid, how to get information on various industries or a description of the space age buildings. When one of the kids noticed that the game posted a suggestion that there should be a transit system, Ian informed him that he didn't need to worry about things like that. After all, with the secret code, there was no need to generate revenue and hence no need to keep one's city happy and well populated. Other children at the club formed an appreciative audience. The adults at the club were in the uneasy position of trying to validate Ian's technical expertise, but wanting to reproduce the quasi-educational urban-planning scenario and not the action-entertainment content of Ian's destructive scenarios. They were foiled by the content of a game that has these capabilities as a hidden interactional resource. The result was that Ian's subjectivity was produced as a counter-cultural subjectivity that subverted educators' expectations but enlisted the support of other kids at the club, who relished this subversion. According to Consolvo (2007), cheats are an example of a particular form of gaming capital that can confer both status and advantage to players. Ian was a political actor who hacked technology to enlist it into his oppositional social network.

Ian's play with *SimCity 2000* makes visible the complicated network of relations that links players, games, and particular contexts of game play. Within the 5thD after-school program, blowing up buildings was a subversive activity, going against the club's educational goals and the game's orientation as it had been widely marketed. Yet the designers of *SimCity 2000* obviously anticipated opportunities for destruction and coded them into the game, following well-established idioms of action gaming. Behind educators' backs, the game designers mobilized a powerful counternarrative that enlisted computer-savvy kids at the 5thD site. The game and the Internet gave Ian access to subcultural but powerful adult communities who provided the resources to validate a subjectivity and practices that worked in opposition to both the club's educational goals and the adults at the site. Ian's play with the alternative functionality of *SimCity* built relationships with other kids as well as with fan communities and game designers. Developer Will Wright and *SimCity* fans thus served as a resource for Ian to produce an alternative subjectivity and social network that could stand up to the club adults' demands. At one level, this case is about familiar antagonisms and fault lines between children and their adult oppressors as children struggle for autonomy within adult-run institutions. In this long-standing power struggle, however, new interlocutors have entered the mix, handing children new resources and sources of solidarity.

In his discussion of games and learning, James Paul Gee (2003) discusses the affinity groups that support social forms of learning through gaming.

Geeked-out gaming groups are a resource for kids to experiment with more adultlike agency and subcultural identification (Ito, Baumer, Bittanti, et al. 2009). Ian was able to tap into this knowledge network and transform the game into a radically different space of possibility. Instead of claiming slivers of time to construct, say, a frivolous freeway before having to start over, he was able to construct and blow up as many fusion plants and large, space-age buildings as he desired. He wholly escaped the subjectivity of a responsible and constructive mayor and instead was able to smuggle in an entertainment-based code, while still passing as a *SimCity 2000* expert at the 5thD. The codes Ian mobilized are not individualistic or antisocial, but rather part of an alternative community of practice, one that includes other kids, fans, and the *SimCity 2000* game developers.

In addition to the different modes of "legitimate" play, *SimCity 2000* embodies hidden functionality accessible by "cheat codes" that a user can type in to change the game parameters. Today it is expected that games will have "cheats" and "Easter eggs," and the original *SimCity* is one of the games that popularized this tendency. An Easter egg, in contrast to other forms of cheats, is defined on one *SimCity* Web page as "a pre-programmed, hidden and undocumented feature inserted by the programmer for their [*sic*] own enjoyment." In the world of computer-game fandom, the meaning of the term *undocumented* is clearly relative, and cheat codes are featured on even the official Maxis Web page, albeit with a disclaimer:

> We do not advocate the use of cheat codes, as playing legitimately will lead to a far more stable and enjoyable city building experience. We are providing these cheat codes solely as a service to you. We will not discuss these codes further on the phone. We do not guarantee that they will work for you. We will not discuss problems with any city in which you have used a code—once you use a code, all bets are off, and you're on your own! (http://www.maxis.com 1998)

Maxis capitalizes on the flexibility of computer technology in catering to heterogeneous users and use situations. Although the dominant marketing pitch focuses on a constructive and educational simulation, the designers and a wired fan community have successfully smuggled in a myriad of alternative readings, including destructive special effects and Easter eggs for localized and customized forms of game play.

In my interview with Wright, he described the cheat codes in *SimCity 2000* as a way to transform the software from a game with a goal structure to a toy, more like a construction kit. In other words, he consciously sees the codes as enabling alternative forms of game play, a move that he called "engineered subversion," the software's ability to encode the conditions of its own undoing. He also described how game producers use cheat codes to generate buzz about a game. "Buzz" is a marketing tool, even if it means subverting the goal orientations and rules embedded in the game design. In other words, cheat codes are a community-building device. Finding and exploiting cheat codes is itself a form of game play for geeked-out players. Wright talked about how in the early days of computer gaming, hackers would race to break the copy protection of a new piece of software and how they were still doing the same today. "They'll break through the code. They'll disassemble the program. It's incredible. We came out with *SimCity 3000*. People within a week had discovered all the cheat codes by digging through the code." Consolvo (2007) sees cheat codes as one component of a broader set of "paratexts" that surround a game's primary text. Cheats, tips, walk-throughs, and reviews are all examples of paratexts that invite player participation in the broader knowledge economy around gaming.

SimCity grows out of and enlists an extremely varied set of producer and consumer communities and embodies these contradictions within its design. The game's position in both educational and entertainment markets, the use of cheat codes and Easter eggs, and a double-talking Web page are indicators of this complexity. These features are common to any media artifact with a large and diverse set of fan communities, but because of *SimCity*'s status as a digital authoring tool, it is an even more flexible object. Players can also buy the *SimCity2000 Urban Renewal Kit*, which allows them to go behind the scenes, design their own buildings and scenarios, and circumvent other game limitations. Hundreds of Web pages, including Maxis's own, publish and distribute the renewal-kit creations made by players. Instead of only interpretive flexibility, consumers are also handed functional flexibility and the ability to author radically different and personalized narrative terms through their game play.

Conclusions: From Engineered Learning to Engineered Subversion

The construction genre exemplifies the ways in which software and gaming are reconfiguring some of the conditions under which kids engage with social and technical ecologies of learning. Just as Papert argued in the early years of educational computing, software provides a unique set of resources for kids to build, author, and exercise agency. The academic and entertainment genres of software exploit new forms of interactivity and multimedia representation afforded by digital media. The construction genre, in addition, exploits computers' capabilities to support a child-centered approach to production and authoring. Rather than offering particular pieces of content to children, software in the construction genre delivers a set of tools and what Bogost (2007) has described as more procedural types of literacy.

The positioning of this genre is not restricted to the relationship between child and machine, but extends to the broader social and political context in which this relationship is embedded. Construction-oriented software positions kids as producers rather than consumers of content and challenges educational approaches that aim to transmit a standardized body of knowledge. Like the use of popular-culture idioms in the entertainment genre, software in the *SimCity 2000* vein hands kids new kinds of resources that can challenge adult authority. It differs, however, from the cultural opposition that entertainment media make in creating a children's culture opposed to adult norms of discipline, health, and work. Authoring tools are practically rather than merely symbolically antiauthoritarian; they shift the control of cultural production, allowing children not only to imagine a world where kids rule, but substantively to participate in the construction of this world. They are, to borrow Kinder's (1991) phrase, playing with power, embodying their interests and personalities in a powerful new computational media.

Given the power of these forms of technology engagement, it is critical to interrogate gender dynamics that associate boys with these more generative and geeky technology uses. The 5thD demonstrated the persistent dominance of boys in displaying technical mastery, but it also proved that when given access and support, girls excel in creating virtual worlds tailored to their own interests. As girls move into gaming and Internet spaces in greater numbers, we can expect that historically girl-centered media-making practices will increasingly intersect with the more geeky technology domains. Although there is little evidence that the gendering of technology is on its way out, we are clearly seeing growing diversity in the gender valences of online media and an opportunity for educators and developers to intervene in these dynamics.

The construction genre exhibits unique dynamics in relation to age boundaries. Because it is not tied to age-graded forms of knowledge, positioning it as a genre of software targeted specifically to children is difficult. The most successful titles in this genre are not age specific, and games such as *SimCity 2000* and digital media authoring tools offer kids an adultlike subjectivity. Authoring software can place children politically and socially in the same arena as adults, particularly because children often have the upper hand in manipulating new computational media. When a piece of technology becomes part of a network of peer information exchange, children's knowledge can quickly supersede that of their parents and teachers, and they look to mixed-age hacker and fan subcultures to extend their expertise. Affinity groups around gaming and other forms of technical knowledge are examples of what, in other work, I have described as a site of interest-driven, peer-based learning (Ito 2007; Ito, Baumer, Bittanti, et al. 2009). These sites are contexts where kids can participate as peers in a set of knowledge networks that are not age specific. Although there is clearly a place for products in this genre to be tailored to young children, these products often support older children's participation in mixed-aged communities of interests.

In the years between the heyday of *SimCity 2000* and our current moment in the midst of a boom in Web 2.0 social media and user-generated content, we have seen a tremendous expansion in the scope of digital media authoring and related social exchange. The kind of niche-knowledge communities that traffick in the cheat codes that Ian gained access to have exploded on the Internet in the years since he participated in the 5thD. Today's children and youth are growing up in a media ecology where producing, modifying, messing around with, customizing, and sharing digital media are part of everyday life (Ito, Baumer, Bittanti, et al. 2009). Unlike the academic and entertainment genres, the construction genre has expanded rather than narrowing and hardening, providing new kinds of opportunities to consider alternative forms of learning environments in a more child-centered mode. Although the construction genre holds out tremendous promise, the lessons we have learned from its historical development are still applicable as our digital media ecology evolves. The struggles between kids and adults over autonomy and agency, the gender dynamics that can exclude girls from certain forms of technical mastery, and the challenge in defining an educational agenda in this genre are still fully in play.

5 Conclusion

The production, distribution, and play of learning software involve the intertwining of different genres, social agendas, and educational philosophies. These dynamics include the negotiation between adults and kids who are performing the academic, entertainment, and construction genres; the ideals of learning, fun, and creativity; and the politics of enrichment, indulgence, and empowerment. The three genres are tied to different social investments: edutainment is a vehicle for producing class and educational distinction; entertainment produces age-cohort identity by creating a space of childhood pleasures defined in opposition to adult disciplines; and construction supports a subjectivity of creative self-actualization tied to technical mastery. All three genres have much deeper historical roots than the relatively recent turn to learning games and children's software. They draw from longstanding discourses in education and middle-class parenting that attempt to transform play into a site of learning, through either a behaviorist or a constructivist model of learning. The education genre butts up against the entrenched idioms and institutions of commercial entertainment that have taken an increasingly stronger hold on childhood peer cultures in the post-TV era. The adult-oriented goals of progress and constructive play are increasingly defined in opposition to children's cultures of repetitive action, fun, and phantasmagoria. In this concluding chapter, I reflect on how the dynamics I have described inform our understanding of structural conditions and historical trends in learning and new media, and how technology participates in social and cultural change.

Underlying my description of the three genres has been the opposition between the cultural categories of education and entertainment. Contemporary childhood in the United States is largely organized by the institutions of school and home, and each institutional setting has certain social imperatives and associated media products. Well-established media industries and infrastructures insist on these distinctions, segment markets and advertising based on the opposition between play and learning, and build on parents and kids' established genre recognitions. Because the marketing of curricular software to parents or of entertainment to children falls into these established genres, it is much more difficult to market and distribute genre-breaking titles. Although many of the early children's software titles such as *Oregon Trail, Carmen San Diego,* and *SimCity* were not as constrained in this way, genres hardened over time as the industry and market for children's software matured. As the development context shifted from a small, experimental research effort to a mainstream commercial enterprise, the founding impetus of educational and cultural reform shifted to an emphasis on existing institutional and market demands. Children's software titles were increasingly polarized between those that promised to further curricular goals and an achievement-oriented identity, on the one side, and those that were meant to be fun and exciting and to compete with television for children's attentions, on the other.

Early edutainment developers hoped to put accessible technical tools in the hands of the disenfranchised, alleviating the oppressiveness of narrow notions of education. Instead, children's software became another site for addressing achievement anxiety in parents and for supporting achievement for children who seem to have been born into success. To have systemic impact on issues such as educational equity, reform efforts that rely on educational media must produce innovative content as well as innovative distribution mechanisms and contexts of reception. In this book, I have argued for how media content is inseparable from the economic and structural conditions in which it is produced and circulates. Social change needs to be pursued at all levels in the circuit of production, distribution, and consumption (Gay, Hall, Janes, et al. 1997). When new technologies go mainstream, they fall victim to their own success, shifting from experimental technologies controlled by innovative pioneers to conduits for existing social structures and norms. The problem of "using games to make learning fun" cannot be addressed simply as a research or software design problem.

Although as researchers we may recognize the learning potential of entertainment media, this recognition alone does not change the structural conditions that insist on the bifurcation between entertainment and education and that correlate only academic content with educational success. The field at times operates on the assumption that the design of games determines learning outcomes, even though we know that these outcomes are realized only through practices of actual play. Further, these practices result in sustained and systemic changes only if our entrenched institutions take them up in ways that reshape existing institutional structures and practices. Although early developers and boosters of learning games hoped they could transform the conditions of education by the persuasiveness of their games alone, history has shown us the problems with underestimating the power of existing institutions and of overestimating a new technology's influence.

As educational researchers, we are part of the same fabric of culture and society in which we seek to intervene, and the field of games and learning reproduces genres that are widely distributed in our culture. We draw from the same repertoire of genres that kids, software developers, and businesspeople mobilize in their practice. Although the field overall shares a sentiment that games can animate the learning process in new and productive ways, a disconnect between different paradigms and genres often muddies our efforts to make progress on shared agendas for research and intervention. We often engage in healthy debate on the explicit theoretical paradigms that motivate our work as behaviorists, constructivists, and sociocultural learning theorists. What we rarely discuss explicitly, however, are the ways in which our research practice is embedded in various institutional contexts with differing political orientations and practical consequences. Even in the case of field research that takes a sociocultural view, most work focuses on the design and content of games and on the localized study of game play in experimental programs. Developers as well as researchers focus on outcomes of learning in individual children and on the effect of specific programs rather than on a more systemic process of institutional and cultural change. As long as we retain the former focus, new games and the learning field will reproduce many of the same problems that the early developers of edutainment encountered. Documenting the learning properties of games and building experimental interventions may support the building of new research fields and agendas, but do little to further the effort to mobilize new technology effectively in the service of broad-based educational reform. Local innovation, whether of software or educational programs. cannot have systemic impact without serious attention to the broader networks of institutional practice and cultural discourses that contextualize these efforts.

Although I have stressed the three genres' conservative tendencies, the circuit of culture I have described also suggests multiple points of negotiation, juncture, and disjuncture as well as productive points for research intervention. The ongoing contestations between genres of participation and genres of representation suggest ways to appropriate and reshape the categories of education and entertainment that have been handed down to us. Titles in the construction genre create productive confusion regarding what we recognize as educational or entertainment. Just as the 5thD complicates our notions of what an educational institution is, games such as *SimCity* have challenged us to consider alternative ways of recognizing learning. Educational institutions continue to have a determining effect on how childhood success and achievement are measured even outside the classroom; the case of edutainment demonstrates the ways in which genres of education migrate and morph beyond the institutional boundary of school. Contexts of play and informal learning, although seemingly marginal to the high-stakes contestations over educational sorting and achievement, are sites that demonstrate the alignments and disjunctures between the cultural and social structures of children's lives. When these sites of reception join hands with innovative software creators and distributors across the circuit of culture, we can begin to imagine alternative genres of media and participation that are both compelling and sustainable.

Local contexts of play are malleable and open to contestation. Kids are political actors who mobilize cultural, technical, and social resources in pursuing status negotiations and in claiming agency, momentarily resisting adults' progress goals, smuggling in forbidden idioms of action entertainment and spectacle, and using adults to support public knowledge and status displays. As technology and cultural trends increasingly support a malleable palette of styles and genres, these opportunities for creative mobilization are expanding for a media- and technology-savvy generation. In the micropolitics of everyday play, the balance of power among children, adults, and software is constantly shifting. A game such as *SimCity 2000* can transform from a site of gleeful destruction and boyish status display to a contemplative site of conversation between an adult and a child about the relation between wealth and crime. One well-timed intervention can tip the scale toward a different genre, a different mode of engagement, a different power dynamic. The subtleties of software design are also highly significant in these micropolitics. Seemingly trivial design decisions such as score-keeping mechanisms or a particular cheat code can inflect play in substantive ways.

These local negotiations on their own have no force to transform existing institutions of schooling, conditions of structural inequity, or the family's resilient cultural scripts. They become part of systemic change only when they link up to institutional and broader cultural trends and changes. The historical trajectory of children's software I have described has continued to unfold. In the late 1990s, as the Internet started making inroads into children's everyday lives, CD-ROMs targeted specifically to children began to lose their luster. Even in the heyday of children's software, only a limited sector of families used these products (Buckingham and Scanlon 2003; Giacquinta, Bauer, and Levin 1993). Today, this market is even smaller. Children's software has a niche in school computer labs, in titles marketed to young children, and in platforms such as LeapPad and Leapster that are designed for early readers. In both the school and the home, the uptake of children's software has hardly been transformative in the ways that early proponents imagined. Research has documented how use of these kinds of titles is limited and reproduces the conditions that already exist in schools and families (Buckingham 2007; Cuban 2001; Seiter 2005, 2007). As Larry Cuban (2001) documents in his study of schools in the Silicon Valley, most schools use the technology in relatively limited ways, often relying on commercial edutainment software, rather than make fundamental changes to pedagogy. The software has been domesticated to the institutional accountabilities of school, fitting the compartmentalized nature of most school schedules and forms of computers use and fitting almost entirely within the genre of academic software that integrates curricular goals with a "sugar coating" of entertainment idioms.

On the other side of the aisle, children's uses of computers in the home are largely oriented toward the entertainment genre. The appeal of software with more explicit educational goals is limited to those families with a more "aspiring" or "anxious" approach to education achievement (Buckingham and Scanlon 2003). In contrast to most families' rejection of school-like software, there is a growing appreciation of popular entertainment as a site of learning and intrinsic motivation. We are seeing shifts in the cultural valences attached to new media as a more permissive approach to parenting, and access to popular entertainment is permitted in many educated families. Surveys of media consumption in the United

States indicate that until the 1990s, higher socioeconomic status and school achievement tended to be associated with less exposure to popular media such as television and video games. In recent years, however, this tendency has begun to disappear, with one 2004 survey of children's media use indicating that college-educated families reported the highest levels of children's media use, followed by families with parents who have a high school education. Those in the middle, families with parents who completed some higher education, had the most limited media exposure (Roberts and Foehr 2008). The rise of more complex forms of media that has come with the digital age and the proliferation of sophisticated new series in television is tied to a cultural shift in how families value and manage both media and their children's exposure to these media. Stephen Johnson's (2005) book *Everything Bad Is Good for You* is an account of these changing cultural winds. His argument for the cognitively complex nature of engagement with contemporary media has been taken up widely as a compelling counterargument to the notion that popular media involves a dumbing down of culture. In other words, his alternative public script about media and learning is challenging some of the existing public scripts.

The growing appreciation for the educative dimensions of popular media is tied to a broader shift in where we look for culture, knowledge, and everyday sociability (Benkler 2006; Jenkins 2006; Varnelis 2008). With the spread of the Internet and low-cost digital authoring tools, kids have a broader social and technological palette to engage in self-authoring and digital media production. With the growing popularity of online journaling, social network sites, game modding, and remix cultures, we are in the midst of a cultural shift that positions digital authoring and publication more centrally in young people's peer cultures (Ito, Baumer, Bittanti, et al. 2009). David Buckingham describes this shift as a "new 'digital divide' between in-school and out-of-school use," which he sees as indicative of "a widening gap between children's everyday 'life worlds' outside school and the emphases of many education systems" (2007, 96). The content and traditional learning formats of schools, particularly when they have to do with new media, is increasingly out of step with the energies that young people are bringing to these new media practices in their social and recreational lives.

If I were to place my bet on a genre of participation that has the potential to transform the systemic conditions of childhood learning, I would pick the construction genre. These geekier and more hacker-oriented forms of play and software are moving from the periphery to the center in this changing new media ecology; they hold out the promise of offering more participatory ways of learning and engaging in media (Ito 2006b, 2008; Jenkins 2006; McGonigal 2006; Ondrejka 2007). This change is interacting with the shifting balance between learning in the home and learning at school, and between entertainment and educational media. As the overall media ecology in which children are immersed shifts, construction-oriented learning games can more effectively function as a scaffold for genres of media engagement and learning that do not rely on an opposition between education and entertainment, learning and play. These new practices challenge our research frameworks and the broader institutional structures of capitalism, entertainment, education, and the family. Unlike existing education practices, the forms of learning attached to popular culture are not dominated by the public institutions of schooling, involve a much more active role for commercial enterprise, and are more deeply embedded in the everyday contexts of the home. In other words, the construction genre is beginning to enlist a diverse range of allies in a wide range of institutional settings.

My argument is not that construction-oriented software, participatory media cultures, and Internet enterprises on their own have the force to transform kids' learning or the shape of educational institutions. Rather, it is that we will begin to see systemic effects only when the different sites of practice link up across the circuit of culture, coalesce into recognizable genres of culture and participation, and become embedded in the structure and practices of institutions. Research has a role to play in these negotiations, not only in documenting these processes, but in giving voice to and reshaping different public scripts and genres that circulate in our everyday lives. Our research practice is part of an ongoing struggle about what forms of learning we value and about the waxing and waning power of different institutions. The transformative effects of new technologies can never be read off the content or design of the technology itself because of the indeterminate nature of these dynamic processes of alliance building. The story of how children's software was incorporated into different genres of practice and institutions and the unfolding history of how these technologies are evolving illustrate these dynamic and highly distributed processes of social and cultural change.

Notes

1. For example, Reed Stevens, Tom Satwicz, and Laurie McCarthy (2007) examine how game play is embedded in the ecology of the home and in everyday play with friends and siblings, Holin Lin (2008) has examined how physical gaming spaces shape game experience and access, and some of my own work has looked at how gaming relates to engagement with other kinds of interest communities and friend-ship networks (Ito and Bittanti 2009). A recent collection of essays, *Gaming Lives in the Twenty-First Century*, edited by Cynthia L. Selfe and Gail E. Hawisher (2007), draws from James Paul Gee's (2003) arguments and documents the ways in which video game play has filtered into young people's learning trajectories and the devel-opment of a wide range of different literacies. Mia Consolvo (2007) looks at player activity in relation to "gaming capital" and broader discourses of cheating and competition in play. Some recent survey work has also documented how game play has become an integral part of social and in some cases civic engagement (Kahne, Middaugh, and Evans 2008). Finally, a growing literature on multiplayer online gaming practice is also examining the broader political, economic, and discursive contexts of gaming practice (eg., Castronova 2006; Dibbell 2006; Steinkuehler 2006; Taylor 2006).

2. Cellular automata are a computation method developed by John Von Neumann and artificial-life scientists in an attempt to model mathematically a self-reproducing automaton (Helmreich 2000).

References

Abu-Lughod, Lila. 1995. The Objects of Soap Opera: Egyptian Television and the Cultural Politics of Modernity. In *Worlds Apart: Modernity through the Prism of the Local,* ed. D. Miller, 190–210. New York: Routledge.

Althusser, Louis. 1969. Ideology and State Apparatuses. In *Lenin and Philosophy and Other Essays,* by Louis Althusser, 127–188. New York: Monthly Review Press.

Ambron, Sueanne. 1989. Introduction. In *Interactive Multimedia: Visions of Multimedia for Developers, Educators, & Information Providers,* ed. S. Ambron and K. Hooper, 1–12. Learning Tomorrow. Redmond, Wash.: Microsoft Press.

Appadurai, Arjun. 1990. Disjuncture and Difference in the Global Cultural Economy. *Public Culture* 2 (2): 1–24.

Appadurai, Arjun. 1986. Introduction: Commodities and the Politics of Value. In *The Social Life of Things: Commodities in Cultural Perspective,* ed. A. Appadurai, 3–63. New York: Cambridge University Press.

Appadurai, Arjun, and Carol Breckenridge. 1988. Why Public Culture. *Public Culture Bulletin* 1 (1): 5–9.

Banet-Weiser, Sarah. 2007. *Kids Rule! Nickelodeon and Consumer Citizenship.* Durham, N.C.: Duke University Press.

Barol, Bill. 1989. Big Fun in a Small Town: Modeling the Perfect City on a Home Computer. *Newsweek,* May 29, 64.

Baym, Nancy K. 2000. *Tune In, Log On: Soaps, Fandom, and Online Community.* Thousand Oaks, Calif.: Sage.

Benkler, Yochai. 2006. *The Wealth of Networks: How Social Production Transforms Markets and Freedom.* New Haven, Conn.: Yale University Press.

Benkler, Yochai. 2000. From Consumers to Users: Shifting the Deeper Structures of Regulation toward Sustainable Commons and User Access. *Federal Communications Law Journal* 52:561–579.

Bird, S. Elizabeth. 2003. *The Audience in Everyday Life: Living in a Media World*. New York: Routledge.

Boellstorff, Tom. 2008. *Coming of Age in* Second Life: *An Anthropologist Explores the Virtually Human*. Princeton, N.J.: Princeton University Press.

Bogost, Ian. 2007. *Persuasive Games: The Expressive Power of Videogames*. Cambridge, Mass.: MIT Press.

Bourdieu, Pierre. 1972. *Outline of a Theory of Practice*. Trans. R. Nice. New York: Cambridge University Press.

Brown, John Seely, Alan Collins, and Paul Duguid. 1989. Situated Cognition and the Culture of Learning. *Educational Researcher* 18:32–37.

Brown, John Seely, and Paul Duguid. 1996. Keeping It Simple. In *Bringing Design to Software*, ed. T. Winograd, 129–145. New York: ACM Press/Addison-Wesley.

Buckingham, David. 2007. *Beyond Technology: Children's Learning in the Age of Digital Culture*. Malden, Mass.: Polity.

Buckingham, David. 1993. *Children Talking Television: The Making of Television Literacy*. New York: RoutledgeFarmer.

Buckingham, David, and Margaret Scanlon. 2003. *Education, Entertainment, and Learning in the Home*. Philadelphia: Open University Press.

Buckingham, David, and Julian Sefton-Green. 2004. Structure, Agency, and Pedagogy in Children's Media Culture. In *Pikachu's Global Adventure: The Rise and Fall of Pokémon*, ed. J. Tobin, 12–33. Durham, N.C.: Duke University Press.

Bury, Rhiannon. 2005. *Cyberspaces of Their Own: Female Fandoms Online*. New York: Peter Lang.

Camp, L. Jean. 1996. We Are Geeks, and We Are Not Guys: The Systers Mailing List. In *Wired_Women: Gender and New Realities in Cyberspace*, ed. L. Cherny and E. R. Weise, 114–125. Seattle: Seal Press.

Cassell, Justine, and Meg Cramer. 2007. High Tech or High Risk: Moral Panics about Girls Online. In *Digital Youth, Innovation, and the Unexpected*, ed. T. McPherson, 53–76. Cambridge, Mass.: MIT Press.

Cassell, Justine, and Henry Jenkins, eds. 1998. *From Barbie to* Mortal Kombat: *Gender and Computer Games*. Cambridge, Mass.: MIT Press.

Cole, Michael. 1997. *Cultural Psychology: A Once and Future Discipline*. Cambridge, Mass.: Harvard University Press.

Cole, Michael, ed. 1994. *Using New Information Technologies in the Creation of Sustainable Afterschool Literacy Activities: From Invention to Maximizing the Potential*.

San Diego: Laboratory of Comparative Human Cognition, University of California at San Diego.

Cole, Michael, and Distributed Literacy Consortium. 2006. *The Fifth Dimension: An After-School Program Built on Diversity.* New York: Russell Sage Foundation.

Cole, Michael, Yrjo Engestrom, and Olga Vasquez, eds. 1997. *Mind, Culture, and Activity: Seminal Papers from the Laboratory of Comparative Human Cognition.* Cambridge, U.K.: Cambridge University Press.

Consolvo, Mia. 2007. *Cheating: Gaining Advantage in Videogames.* Cambridge, Mass.: MIT Press.

A Conversation with Jan Davidson. 1997. *Children's Software Review* 5 (3):25–26.

Castronova, Edward. 2006. *Synthetic Worlds: The Business and Culture of Online Games.* Chicago: University of Chicago Press.

Coppa, Francesca. 2008. Women, *Star Trek,* and the Early Development of Fannish Vidding. *Transformative Works and Cultures* 1. http://journal.transformativeworks .org/index.php/twc/article/view/44.

Cosaro, William. 1997. *The Sociology of Childhood.* London: Pine Forge Press.

Cross, Gary. 1997. *Kids' Stuff: Toys and the Changing World of American Childhood.* Cambridge, Mass.: Harvard University Press.

Cuban, Larry. 2001. *Oversold & Underused: Computers in the Classroom.* Cambridge, Mass.: Harvard University Press.

Dargahi, Nick, and Michael Bremer. 1995. SimCity^{TM}: *Power, Politics, and Planning.* Rev. ed. Rocklin, Calif.: Prima.

Darlin, Damon. 1994. Early Bird versus the Flock. *Forbes* 154 (2): 298–299.

Debord, Guy. 1995. *Society of the Spectacle.* Detroit: Black & Red.

Dibbell, Julian. 2006. *Play Money: Or How I Quit My Day Job and Made Millions Trading Virtual Loot.* New York: Basic Books.

Edwards, Lynn Y. 2005. Victims, Villains, and Vixens. In *Girl Wide Web: Girls, the Internet, and the Negotiation of Identity,* ed. S. R. Mazzarella, 13–30. New York: Peter Lang.

Edwards, Paul. 1995. From "Impact" to Social Process: Computers in Society and Culture. In *Handbook of Science and Technology Studies,* ed. S. Jasanoff, G. E. Markle, J. C. Petersen, and T. Pinch, 257–285. Thousand Oaks, Calif.: Sage.

Eisler, Leslie. 1991. Learning to Save the Environment. *Technology and Learning* (March): 18–23.

Engenfeldt-Nielsen, Simon. 2006. Overview of Research on the Educational Use of Video Games. *Digital Kompetanse* 1 (3): 184–213.

Engestrom, Yrjo. 1993. Developmental Studies of Work as a Testbench of Activity Theory: The Case of Primary Care Medical Practice. In *Understanding Practice: Perspectives on Activity and Context,* ed. S. Chaiklin and J. Lave, 64–103. New York: Cambridge University Press.

Escobar, Arturo. 1994. Welcome to Cyberia: Notes on the Anthropology of Cyberculture. *Current Anthropology* 35 (3): 211–232.

Fischer, Michael M. J. 1991. Anthropology as Cultural Critique: Inserts for the 1990s, Cultural Studies of Science, Visual-Virtual Realities, and Post-trauma Politics. *Cultural Anthropology* 6 (4): 525–537.

Foucault, Michel. 1978. *The History of Sexuality: An Introduction.* Vol. 1. Trans. R. Hurley. New York: Vintage Books.

Frank, Thomas. 1997. *The Conquest of Cool: Business Culture, Counterculture, and the Rise of Hip Consumerism.* Chicago: University of Chicago Press.

Gay, Paul du, Stuart Hall, Linda Janes, Hugh Mackay, and Keith Negus. 1997. *Doing Cultural Studies: The Story of the Sony Walkman.* Thousand Oaks, Calif.: Sage.

Gee, James Paul. 2003. *What Video Games Have to Teach Us about Learning and Literacy.* New York: Palgrave Macmillan.

Giacquinta, Joseph B., Jo Anne Bauer, and Jane E. Levin. 1993. *Beyond Technology's Promise: An Examination of Children's Educational Computing at Home.* Cambridge, U.K.: Cambridge University Press.

Giddens, Anthony. 1986. *The Constitution of Society: Outline of the Theory of Structuration.* Berkeley and Los Angeles: University of California Press.

Goldman, Shelley, and Raymond McDermott. 1987. The Culture of Competition in American Schools. In *Education and Cultural Process: Anthropological Approaches,* 2d ed., ed. G. Spindler, 282–300. Prospect Heights, Ill.: Waveland Press.

Gunter, Barrie. 1998. *The Effects of Video Games on Children: The Myth Unmasked.* Sheffield, U.K.: Sheffield Academic Press.

Gupta, Akhil. 1995. Blurred Boundaries: The Discourse of Corruption, the Culture of Politics, and the Imagined State. *American Ethnologist* 22 (2): 375–403.

Gupta, Akhil, and James Ferguson. 1992. Space, Identity, and the Politics of Difference. *Cultural Anthropology* 7 (1): 6–23.

Harel, Idit. 1991. *Children Designers: Interdisciplinary Constructions for Learning and Knowing Mathematics in a Computer-Rich School.* Norwood, N.J.: Ablex.

Helft, Miguel. 2008. LeapFrog Hopes for Next Hit with Interactive Reading Toy. *New York Times,* January 28, http://www.nytimes.com/2008/01/28/technology/28leapfrog. html?fta-y.

Helmreich, Stefan Gordon. 2000. *Silicon Second Nature: Culturing Artificial Life in a Digital World.* Berkeley: University of California Press.

Hine, Christine. 2000. *Virtual Ethnography.* London: Sage.

Holloway, Sarah L., and Gill Valentine. 2001. *Cyberkids: Children in the Information Age.* New York: Routledge.

Hooper, Paula K. 2007. Looking BK and Moving FD: Toward a Sociocultural Lens on Learning with Programmable Media. In *Digital Youth, Innovation, and the Unexpected,* ed. T. McPherson, 123–142. Cambridge, Mass.: MIT Press.

Hoover, Stewart M., Lynn Shofield Clark, and Diane F. Alters. 2004. *Media, Home, and Family.* New York: Routledge.

Hutchins, Edwin. 1995. *Cognition in the Wild.* Cambridge, Mass.: MIT Press.

Ito, Mizuko. 2008. Mobilizing the Imagination in Everyday Play: The Case of Japanese Media Mixes. In *The International Handbook of Children, Media, and Culture,* ed. K. Drotner and S. Livingstone, 397–412. Thousand Oaks, Calif.: Sage.

Ito, Mizuko. 2007. Technologies of the Childhood Imagination: *Yu-Gi-Oh!* Media Mixes, and Everyday Cultural Production. In *Structures of Participation in Digital Culture,* ed. J. Karaganis, 88–110. New York: Social Science Research Council.

Ito, Mizuko. 2006a. Interaction, Collusion, and the Human-Machine Interface. In *International Handbook of Virtual Learning Environments,* ed. J. Weiss, J. Nolan, and P. Trifonas, 221–240. Norwell, Mass.: Kluwer.

Ito, Mizuko. 2006b. Japanese Media Mixes and Amateur Cultural Exchange. In *Digital Generations,* ed. D. Buckingham and R. Willett, 49–66. Hillsdale, N.J.: Lawrence Erlbaum.

Ito, Mizuko, Sonja Baumer, Matteo Bittanti, danah boyd, Rachel Cody, Becky Herr, Heather A. Horst, Patricia G. Lange, Dilan Mahendran, Katynka Martinez, C. J. Pascoe, Dan Perkel, Laura Robinson, Christo Sims, and Lisa Tripp (with Judd Antin, Megan Finn, Arthur Law, Annie Manion, Sarai Mitnick, and Dan Schlossberg and Sarita Yardi). 2009. *Hanging Out, Messing Around, and Geeking Out: Living and Learning with New Media.* Cambridge, Mass.: MIT Press.

Ito, Mizuko, and Matteo Bittanti. 2009. Gaming. In *Hanging Out, Messing Around, and Geeking Out: Living and Learning with New Media,* by Mizuko Ito, Sonja Baumer, Matteo Bittanti, Danah Boyd, Rachel Cody, Becky Herr, Heather A. Horst, Patricia G. Lange, Dilan Mahendran, Katynka Martinez, C. J. Pascoe, Dan Perkel, Laura Robinson, Christo Sims, and Lisa Tripp (with Judd Antin, Megan Finn,

Arthur Law, Annie Manion, Sarai Mitnick and Dan Schlossberg and Sarita Yardi). Cambridge, Mass.: MIT Press.

Jacobson, Pat. 1992. Save the Cities! *SimCity* in Grades 2–5. *The Computing Teacher* (October): 14–15.

James, Allison, Chris Jenks, and Alan Prout, eds. 1998. *Theorizing Childhood.* New York: Teachers College Press.

James, Allison, and Alan Prout, eds. 1997. *Constructing and Reconstructing Childhood: Contemporary Issues in the Sociological Study of Childhood.* Philadelphia: Routledge Farmer.

Jenkins, Henry. 2006. *Convergence Culture: Where Old and New Media Collide.* New York: New York University Press.

Jenkins, Henry. 1998. Introduction: Childhood Innocence and Other Modern Myths. In *The Children's Culture Reader,* ed. H. Jenkins, 1–37. New York: New York University Press.

Jenkins, Henry. 1992. *Textual Poachers: Television Fans and Participatory Culture.* New York: Routledge.

Johnson, Steven. 2005. *Everything Bad Is Good for You.* New York: Riverhead Books.

Kafai, Yasmin B. 1995. *Minds in Play: Computer Game Design as a Context for Children's Learning.* Hillsdale, N.J.: Lawrence Erlbaum.

Kafai, Yasmin B., Carrie Heeter, Jill Denner, and Jennifer Y. Sun, eds. 2008. *Beyond Barbie and Mortal Kombat: New Perspectives on Gender and Gaming.* Cambridge, Mass.: MIT Press.

Kahne, Joseph, Ellen Middaugh, and Chris Evans. 2008. *The Civic Potential of Video Games.* Oakland, Calif.: Civic Engagement Research Group.

Kay, Alan C. 1991. Computers, Networks, and Education. *Scientific American* (September 1991): 138–148.

Kearney, Mary Celeste. 2006. *Girls Make Media.* New York: Routledge.

Kendall, Lori. 2002. *Hanging Out in the Virtual Pub: Masculinities and Relationships Online.* Berkeley and Los Angeles: University of California Press.

Kinder, Marsha. 1999. Kids' Media Culture: An Introduction. In *Kids' Media Culture,* ed. M. Kinder, 1–12. Durham, N.C.: Duke University Press.

Kinder, Marsha. 1991. *Playing with Power in Movies, Television, and Video Games.* Berkeley and Los Angeles: University of California Press.

Kline, Stephen. 1993. *Out of the Garden: Toys and Children's Culture in the Age of TV Marketing.* New York: Verso.

Kline, Stephen, Nick Dyer-Witherford, and Greig de Peuter. 2003. *Digital Play: The Interaction of Technology, Culture, and Marketing.* Montreal: McGill-Queen's University Press.

Kozulin, Alex. 1986. Vygostky in Context. In *Thought and Language,* ed. A. Kozulin, xi–lvi. Cambridge, Mass.: MIT Press.

Kutner, Lawrence, and Cheryl K. Olson. 2008. *Grand Theft Childhood: The Surprising Truth about Violent Video Games and What Parents Can Do.* New York: Simon & Schuster.

Lareau, Annette. 2003. *Unequal Childhoods: Class, Race, and Family Life.* Berkeley and Los Angeles: University of California Press.

Lave, Jean. 1993. The Practice of Learning. In *Understanding Practice: Perspectives on Activity and Practice,* ed. S. Chaiklin and J. Lave, 3–34. New York: Cambridge University Press.

Lave, Jean. 1988. *Cognition in Practice.* New York: Cambridge University Press.

Lave, Jean, and Etienne Wenger. 1991. *Situated Learning: Legitimate Peripheral Participation.* New York: Cambridge University Press.

Lessig, Lawrence. 1999. *Code and Other Laws of Cyberspace.* New York: Basic Books.

Levy, Steven. 1994. *Hackers: Heroes of the Computer Revolution.* New York: Anchor Press.

Lewis, Peter H. 1992. Programs to Hold a Child's Interest. *New York Times,* June 2, 2.

Lin, Holin. 2008. Body, Space, and Gendered Gaming Experiences: A Cultural Geography of Homes, Cybercafés, and Dormitories. In *Beyond Barbie and* Mortal Kombat: *New Perspectives on Gender and Gaming,* ed. Y. B. Kafai, C. Heeter, J. Denner, and J. Y. Sun, 67–82. Cambridge, Mass.: MIT Press.

Livingstone, Sonia. 2002. *Young People and New Media.* London: Sage.

Lowood, Henry. 2007. Found Technology: Players as Innovators in the Making of Machinima. In *Digital Youth, Innovation, and the Unexpected,* ed. T. McPherson, 165–196. Cambridge, Mass.: MIT Press.

MacBeth, Douglas, and Michael Lynch. 1997. Telewitnessing: Elementary Spectacles of Science Education. Paper presented at the annual meeting for the Society for the Social Studies of Science, October 23–26, Tucson, Arizona.

Mankekar, Purnima. 1999. *Screening Culture, Viewing Politics: An Ethnography of Television, Womanhood, and Nation in Postcolonial India.* Durham, N.C.: Duke University Press.

Marcus, George E. 1998. *Ethnography through Thick and Thin.* Princeton, N.J.: Princeton University Press.

Marcus, George E., ed. 1996. *Connected: Engagements with Media.* Vol. 3. Chicago: University of Chicago Press.

Marcus, George. 1995. Ethnography in/of the World System: The Emergence of Multi-sited Ethnography. *Annual Review of Anthropology* 24:95–117.

Martin, Emily. 1994. *Flexible Bodies: Tracking Immunity in American Culture—From the Days of Polio to the Age of AIDs.* Boston: Beacon Press.

Mazzarella, Sharon R., ed. 2005. *Girl Wide Web: Girls, the Internet, and the Negotiation of Identity.* New York: Peter Lang.

McDermott, Raymond, Kenneth Gospodinoff, and Jeffrey Aron. 1978. Criteria for an Ethnographically Adequate Description of Concerted Activities and Their Contexts. *Semiotica* 24:245–275.

McGonigal, Jane. 2006. The Puppet Master Problem: Design for Real-World, Mission-Based Gaming. In *Second Person: Role-Playing and Story in Games and Playable Media,* ed. P. Harrigan and N. Wardrip-Fruin, 251–264. Cambridge, Mass.: MIT Press.

Michael, David, and Sande Chen. 2006. *Serious Games: Games That Educate, Train, and Inform.* Boston: Thomson Course Technology.

Miller, Daniel. 1997. *Capitalism: An Ethnographic Approach.* New York: Berg.

Miller, Daniel, and Don Slater. 2000. *The Internet: An Ethnographic Approach.* New York: Berg.

Miranker, Cathy, and Alison Elliot. 1996. *Great Software for Kids and Parents.* Foster City, Calif.: IDG Books.

Morley, David. 1992. *Television, Audiences, and Cultural Studies.* New York: Routledge.

Nader, Laura. [1969] 1972. Up the Anthropologist—Perspectives Gained from Studying Up. In *Reinventing Anthropology,* ed. D. Hymes, 284–311. New York: Pantheon Books.

Nelson, Theodore H. 1974. *Computer Lib/Dream Machines.* N.p.: Self-published.

Nicolopolou, Ageliki., and Michael Cole. 1992. The Fifth Dimension, Its Playworld, and Its Institutional Contexts: The Generation and Transmission of Shared Knowledge in the Culture of Collaborative Learning. In *Contexts for Learning: Sociocultural Dynamics in Children's Development,* ed. E. A. Forman, N. Minnick, and C. A. Stone, 283–314. New York: Oxford University Press.

Ondrejka, Cory. 2007. Education Unleashed: Participatory Culture, Education, and Innovation in *Second Life.* In *The Ecology of Games: Connecting Youth, Games, and Learning,* ed. K. Salen, 229–252. Cambridge, Mass.: MIT Press.

Palfrey, John, and Urs Gasser. 2008. *Born Digital: Understanding the First Generation of Digital Natives*. New York: Basic Books.

Papert, Seymour. 1996. *The Connected Family: Bridging the Digital Generation Gap*. Atlanta: Longstreet Press.

Papert, Seymour. 1993. *The Children's Machine: Rethinking School in the Age of the Computer*. New York: Basic Books.

Papert, Seymour. 1980. *Mindstorms: Children, Computers, and Powerful Ideas*. New York: Basic Books.

Paul, Ronald H. 1991. Finally, a Good Way to Teach City Government! *Social Studies* (July–August): 165–166.

Peirce, Neal R. 1994. Kids Design the Darnedest Cities. *National Journal* (May 21): 1204.

Penley, Constance. 1997. *NASA/Trek: Popular Science and Sex in America*. New York: Verso.

Pham, Alex. 2002. Educational Software Finds Itself at the Back of the Class for Kids' Attention. *Los Angeles Times*, December 16, C-1

Piestrup, Ann. 1984. Game Sets and Builders: Graphics-Based Learning Software. *Byte* 9 (6): 215–219.

Pinch, Trevor F., and Wiebe E. Bijker. 1987. The Social Construction of Facts and Artifacts: Or How the Sociology of Science and the Sociology of Technology Might Benefit Each Other. In *The Social Construction of Technological Systems*, ed. W. E. Bijker, T. P. Hughes, and T. F. Pinch, 17–50. Cambridge, Mass.: MIT Press.

Prensky, Mark. 2001. *Digital Game-Based Learning*. St. Paul, Minn.: Paragon House.

Prensky, Mark. 2006. *"Don't Bother Me Mom–I'm Learning."* St. Paul, Minn.: Paragon House.

Radway, Janice A. 1991. *Reading the Romance: Women, Patriarchy, and Popular Literature*. Chapel Hill: University of North Carolina Press.

Resnick, Mitchel. 2006. Computer as Paintbrush: Technology, Play, and the Creative Society. In *Play = Learning: How Play Motivates and Enhances Children's Cognitive and Social-Emotional Growth*, ed. D. Singer, R. Golikoff, and K. Hirsh-Pasek, 192–208. Oxford, U.K.: Oxford University Press.

Richtel, Matt. 2005. Once a Booming Market, Educational Software for the PC Takes a Nose Dive. *New York Times*, August 22, http://www.nytimes.com/2005/08/22/technology/22soft.html.

Rieber, Loyd P., Nancy Luke, and Jan Smith. 1998. Project KID DESIGNER: Constructivism at Work through Play. *Meridian: A Middle School Computer Technologies Journal* 1 (1): 1–19.

Roberts, Donald F. and Ulla G. Foehr. 2008. Trends in Media Use. *Children and Electronic Media.* 18 (1): 39–62.

Roberts, Donald F., Ulla G. Foehr, and Victoria Rideout. 2005. *Generation M: Media in the Lives of 8–18 Year-Olds.* Menlo Park, Calif.: Kaiser Family Foundation.

Ruocco, Jennifer, and Donald A. Dyson. 1996. CD-ROMs on Sexuality-Related Issues. *SIECUS Report* 25:16–20.

Russell, Adrienne, Mizuko Ito, Todd Richmond, and Mark Tuters. 2008. Culture: Networked Public Culture. In *Networked Publics,* ed. K. Varnelis, 43–76. Cambridge, Mass.: MIT Press.

Seiter, Ellen. 2007. Practicing at Home: Computers, Pianos, and Cultural Capital. In *Digital Youth, Innovation, and the Unexpected,* ed. T. McPherson, 27–52. Cambridge, Mass.: MIT Press.

Seiter, Ellen. 2005. *The Internet Playground: Children's Access, Entertainment, and Miseducation.* New York: Peter Lang.

Seiter, Ellen. 1999. Power Rangers at Preschool: Negotiating Media in Child Care Settings. In *Kids' Media Culture,* ed. M. Kinder, 239–262. Durham, N.C.: Duke University Press.

Seiter, Ellen. 1995. *Sold Separately: Parents and Children in Consumer Culture.* New Brunswick, N.J.: Rutgers University Press.

Selfe, Cynthia L., and Gail E. Hawisher, eds. 2007. *Gaming Lives in the Twenty-First Century: Literate Connections.* New York: Palgrave Macmillan.

Shaffer, David Williamson. 2006. *How Computer Games Help Children Learn.* New York: Palgrave Macmillan.

Silverstone, Roger, Eric Hirsch, and David Morley. 1992. Information and Communication Technologies and the Moral Economy of the Household. In *Consuming Technologies: Media and Information in Domestic Spaces,* ed. R. Silverstone and E. Hirsch, 9–17. New York: Routledge.

Squire, Kurt. 2007. Open-Ended Video Games: A Model for Developing Learning for an Interactive Age. In *The Ecology of Games: Connecting Youth, Games, and Learning,* ed. K. Salen, 167–198. Cambridge, Mass.: MIT Press.

Steinkuehler, Constance. 2006. The Mangle of Play. *Games and Culture* 1 (3): 199–213.

Stevens, Reed, Tom Satwicz, and Laurie McCarthy. 2007. In-Game, In-Room, In-World: Reconnecting Video Game Play to the Rest of Kids' Lives. In *The Ecology of Games: Connecting Youth, Games, and Learning,* ed. K. Salen, 41–66. Cambridge, Mass.: MIT Press.

Suchman, Lucy. 1987. *Plans and Situated Actions: The Problem of Human/Machine Communication.* New York: Cambridge University Press.

Sutton-Smith, Brian. 1997. *The Ambiguity of Play.* Cambridge, Mass.: Harvard University Press.

Tanner, Clive. 1993. *SimCity* in the Classroom. *Classroom: The Magazine for Teachers* 13:37–39.

Tapscott, Don. 1998. *Growing Up Digital: The Rise of the Net Generation.* New York: McGraw-Hill.

Taylor, T. L. 2008. Becoming a Player: Networks, Structure, and Imagined Futures. In *Beyond Barbie and* Mortal Kombat: *New Perspectives on Gender and Gaming,* ed. Y. B. Kafai, C. Heeter, J. Denner, and J. Y. Sun, 51–66. Cambridge, Mass.: MIT Press.

Taylor, T. L. 2006. *Play between Worlds: Exploring Online Game Culture.* Cambridge, Mass.: MIT Press.

Thorne, Barrie. 1993. *Gender Play: Girls and Boys in School.* New Brunswick, N.J.: Rutgers University Press.

Tobin, Joseph, ed. 2004. *Pikachu's Global Adventure: The Rise and Fall of Pokémon.* Durham, N.C.: Duke University Press.

Turkle, Sherry. 1995. *Life on the Screen: Identity in the Age of the Internet.* New York: Simon & Schuster.

Turkle, Sherry. 1984. *The Second Self: Computers and the Human Spirit.* New York: Touchstone.

Varenne, Hervé, Shelley Goldman, and Ray McDermott. 1998. Racing in Place. In *Successful Failure: The School America Builds,* ed. H. Varenne and R. McDermott, 106–128. Boulder, Colo.: Westview Press.

Varenne, Hervé, and Ray McDermott, eds. 1998. *Successful Failure: The School America Builds.* Boulder, Colo.: Westview Press.

Varnelis, Kazys, ed. 2008. *Networked Publics.* Cambridge, Mass.: MIT Press.

Vasquez, Olga A., Lucinda Pease-Alvarez, and Sheila M. Shannon. 1994. *Pushing Boundaries: Language and Culture in a Mexicano Community.* Cambridge, U.K.: Cambridge University Press.

Vygotsky, Lev Semonovich. 1987. *The Collected Works of L. S. Vygotsky.* Trans. N. Minick. New York: Springer.

Wajcman, Judy. 1991. *Feminism Confronts Technology.* University Park: University of Pennsylvania.

Warschauer, Mark. 2003. *Technology and Social Inclusion: Rethinking the Digital Divide.* Cambridge, Mass.: MIT Press.

Index

Educational content
commodification of, 141
vs. entertainment, 116–121, 134
as selling point, 101–103
Educational games, 6, 166
Educational philosophies, 187
"whole language," 93
Educational reform, 188
Educational research, 30, 36–38, 92–93
shift to marketing, 188
social context of, 189–190
Educational software
labeling of, 156
marketing of, 44
vs. video games, 35
Educational tools, programming as, 143
Educational vs. action content, 181
Education Development Center, 93
Edutainment, 2, 5–6, 109, 141, 151, 187
arguments for, 7–8
vs. authoring genre, 151
criticism of software, 143
definition of, 30
developers, goals of, 188, 189
features, of *MSBEHB*, 109
heyday of, 100
history of, 26, 31–43
link to class, academic achievement, 187
marginalization of, 191
market for, 83–84
programs, success of, 150–151
software, 143, 144
Edwards, Lynn Y., 1, 149
Edwards, Paul, 11
Effects, antisocial, of game playing, 127
Effects, media, 13
Effects, special, 177
auditory/sound, 114, 118, 120, 126–131, 159, 162–163

engagement with, 121–131
interactional, engagement with, 126–131
logic of, 127, 129
in *MSBEHB*, 114, 116–121, 129–130
pleasure in, 126–128, 129–130
of *SimCity 2000*, 159, 162–163
visual, engagement with, 121–126
Eisler, Leslie, 156–159
Electronic Arts, 159
Elliot, Alison, 100
Empowerment, 35–36, 187
of hackers, 145, 146
packaging of, 155–156
EncycloAlmanacTionaryOgraphy, 61
Engagement, 126
with academic content, 82, 83
with auditory special effects, 126–131
civil, and game play, 195n1
as "fun," 131, 134
"geeked out," with *SimCity 2000*, 170
girls', in *SimCity 2000*, 170
with interactional special effects, 126–131
with *MSBEHB*, 116–121
as obsession, 178
social, and game play, 195n1
"soft," 147
with special effects, 121–131, 177
with spectacle, 134
with visual special effects, 121–126
Engenfeldt-Nielsen, Simon, 6, 46
Engestrom, Yrjo, 18, 19
"Engineered subversion," 183
Enlistment, of peers, 120–121, 131, 132, 136
Enrichment, childhood, 100
Entertainment genre of participation, 26, 85–141. *See also* Fun
vs. authoring, 151